BARRON'S BACK TO BASICS SERIES

MATHEMATICS
VOLUME 2
Fractions and Decimals

by Edward Williams

Assistant Principal-Supervision, Mathematics Department
Washington Irving High School, New York City

BARRON'S

BARRON'S EDUCATIONAL SERIES, INC.

To Marilyn for her
infinite patience
and encouragement

© 1979 Barron's Educational Series, Inc.

A large portion of this was previously published as part
of *Barron's How to Prepare for High School
Equivalency Examinations—Beginning Preparation
in Mathematics,* ® 1977.

All inquiries should be addressed to:
Barron's Educational Series, Inc.
250 Wireless Boulevard
Hauppauge, New York 11788

Library of Congress Catalog Card No. 78-21490

International Standard Book No. 0-8120-692-5

Library of Congress Cataloging in Publication Data
Williams, Edward, 1926-
  Mathematics.
  (Barron's back to basics series)
  CONTENTS: 1. From addition to divsion.   2. Im-
proving your skills with fractions.   3. Percents and
other matters.
  1. Mathematics — 1961-   (1. Mathematics)
I. Title.
QA39.2.W5583   510   78-21490
ISBN 0-8120-0692-5

PRINTED IN THE UNITED STATES OF AMERICA

456   510   15

Contents

preface

In this series we plan to help students improve their skills in mathematics. You know from daily living that mathematics must be used in many ways. Basic arithmetic — adding, subtracting, multiplying, and dividing — applies to topics such as buying and selling, wages, banking, insurance, taxes, and measurement. As an adult you will have to deal with these problems and you will need more advanced skills.

In this series of books we will provide the explanations and practice for these mathematical skills. We also show you how to study topics of mathematics in a logical and practical order. You should try the test at the beginning of each chapter to see if you understand the ideas well or need to review them. The series is set up so you can move at your own pace.

Barron's Back to Basics Series — Mathematics — will help you advance to high school level work. Topics near the end of the third volume will be more advanced, but you will be ready for them. Many others have used these skills to move ahead to better grades and an improved understanding of mathematics. Your plans for yourself are important; we want to help you make them come true.

I would like to acknowledge the assistance of the many people who helped to shape this series. My thanks go to Dr. Eugene Farley, to Lester Schlumpf, and to Linda Bucholtz Ross.

introduction

Barron's Back to Basics Series — Mathematics reviews and explains basic mathematics. For those of you who want to improve your chances for a better job or to be better informed when you buy something or do business with someone, mathematical skills are important. With this series, you can move from beginning level mathematics to high school level, at your own speed. You can use this series on your own or in class with a teacher to help.

What This Book Can Do

Barron's Back to Basics Series — Mathematics will try to:
1. Review and explain the basic ideas and skills of beginning mathematics.
2. Give many examples to practice the ideas and skills.
3. Increase your confidence to try more advanced material.
4. Encourage you to do better on your mathematics tests in school.

You could summarize the purposes of this book as building your skills and confidence.

How The Series Is Organized

We emphasize basic mathematics — addition, subtraction, multiplication, and division. Each of these is applied to whole numbers, fractions, decimals, and measurement. Later books introduce topics you will cover more at high school level. The ideas and skills are related to problems of everyday life at work or at home.

Each chapter begins with a test so that you can check whether you know and understand the ideas. This is called a pretest because it comes before the chapter work. If you do well on the pretest you can save time by skipping the material in that chapter. If you find you need some help, you have the advantage of knowing which ideas are not clear to you. Then you can read the chapter with a specific purpose. Each topic and each method will be carefully explained. Practice examples are given to help you learn the idea.

After studying each chapter you can check how well you have learned by taking the *posttest*, so named because it comes *after* the chapter work. If you do well on the posttest, you can feel confident about moving to the next chapter. You may find some problems that you cannot do. To help you, there will be a section at the end of the chapter which guides you to the parts you need to study again.

The standard for doing well varies for each test. You can progress as fast as you can study and understand. Answers are given so that you may check whether you did the problems correctly. Obviously, there is no value to looking at the answer unless you have worked the problem. You want to know *how* to do the problems and to learn the technique so that you can use it on similar problems.

For additional interest and a little relaxation, there are some mathematical games and puzzles scattered throughout the book. These also give you practice in the basic mathematical skills.

How To Use This Book

Each individual studies in his or her own way. However, the series is planned in such a way that the following suggestions should work best for you:

1. Read the brief introduction to the topic at the beginning of a chapter.
2. Take the pre-test. Check your answers to decide whether
   a. You need to study that chapter, or
   b. You can move to the next chapter.
3. If you are not sure, study the material in the chapter by
   a. Reading the explanations and
   b. Working the practice examples.
      (*Note – Practice examples are necessary to gain good control over the material. Do not skip the practice until you are sure you do the problems well.*)
4. Take the posttest at the end of the chapter to see how well you have learned.
5. Stress explanations and methods.
   (*Note – You are urged to spend most of your time on **explanations**. The **how** and **why** of mathematics, or the **methods** and the **reasons**, are the most important parts. Once you understand the **reason** for something and the **method** or way to do it, you can apply the knowledge to many other problems.*)

reference tables

United States System Of Weights And Measures

Units of Length

12 inches (in.) = 1 foot (ft.)
3 feet (ft.) = 1 yard (yd.)
1,760 yards (yd.) = 1 mile (mi.)

Units of Capacity

16 fluid ounces (fl. oz.) = 1 pint (pt.)
2 pints (pt.) = 1 quart (qt.)
4 quarts (qt.) = 1 gallon (gal.)

Units of Weight

16 ounces (oz.) = 1 pound (lb.)
2,000 pounds (lb.) = 1 ton (T.)

Units of Time

60 seconds (sec.) = 1 minute (min.)
60 minutes (min.) = 1 hour (hr.)
24 hours (hr.) = 1 day (da.)
7 days (da.) = 1 week (wk.)
52 weeks (wk.)   1 year (yr.)

($\approx$ means approximately equal to)

Metric System Of Weights And Measures

Units of Length

10 millimeters (mm) = 1 centimeter (cm)
10 centimeters (cm) = 1 decimeter (dm)
10 decimeters (dm) = 1 meter (m)
1,000 meters (m) = 1 kilometer (km)

Units of Capacity

1,000 milliliters (ml) = 1 liter (l)
1,000 liters (l) = 1 kiloliter (kl)

Units of Weight

1,000 milligrams (mg) = 1 gram (gm)
1,000 grams (gm) = 1 kilogram (kg)

Metric and United States Equivalents (approximate)

1 inch = 2.54 centimeters
1 meter = 39.37 inches
1 kilometer = .62 mile
1 liter = 1.1 quarts
1 gram = .034 ounces
1 kilogram = 2.2 pounds

6 the important parts - fractions

Everyone has the need to know and to use fractions. You sometimes buy portions of things rather than the whole things. You may buy part of a pound rather than the whole pound. You may measure part of a cup rather than the whole cup. You may purchase part of a yard of material, not the whole yard. These are only a few instances in which you need to use fractions. You can probably think of many more.

The pretest which follows can show how well you understand fractions.

See What You Know And Remember — Pretest 6

Try your past skills again on this test. Do as many problems as you can. Write each answer in the space provided. Simplify the answer if possible.

1. Using this diagram, write, as a fraction, the portion of the rectangle that has been shaded. _____

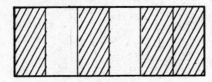

2. If I have 8 items and eat 2 of them, represent, as a fraction, the portion I have eaten.

   _____

3. A fraction consists of two parts, a *numerator* and a *denominator*. In the fraction $\frac{3}{16}$, name the numerator and the denominator.

   3 is the

   _____

   16 is the

   _____

4. Identify each fraction as a *proper* or *improper* fraction.

   a. $\frac{3}{2}$

   _____

   b. $\frac{2}{5}$

   _____

   c. $\frac{4}{4}$

   _____

5. A line is divided into 12 equal segments. What is the name given to one of these parts?

   _____

6. A man has 5 items. If he sells 2 of them, what fractional part remains?

   _____

7. In a classroom of 27 adults, 17 are men.

   a. What fraction of the class is men?

   _____

   b. What fraction of the class is women?

   _____

8. Write the fraction $\frac{12}{32}$ in its simplest form.

   _____

9. To change $\frac{2}{7}$ to an equivalent fraction with a new denominator of 28 you must multiply the numerator and the denominator by a certain number. What number is it?

   _____

10. Change the improper fraction $\frac{14}{5}$ to a mixed number. _____

11. Change the mixed number $3\frac{1}{3}$ to an improper fraction. _____

12. There are approximately 10 hours of daylight each day during the month of October. There are 24 hours in a day.

    a. What fraction of the day is daylight? _____

    b. What fraction of the day is night? _____

13. Find the lowest common denominator for these fractions:

$$\frac{1}{6}, \qquad \frac{2}{3}, \qquad \frac{3}{4}$$

_____

14. Which is the larger of the two fractions:

$$\frac{7}{8} \quad \text{or} \quad \frac{11}{12}$$

_____

15. Arrange these three fractions in order of size, starting with the smallest one first.

$$\frac{2}{3}, \qquad \frac{5}{12}, \qquad \frac{1}{2}$$

_____

Now turn to the end of the chapter to check your answers. Add up all that you had correct. Count by the number of separate answers, not by the number of questions. In this pretest there were 15 questions but 20 separate answers.

A Score of	Means That You
18–20	Did very well. You can move to Chapter 7.
16–17	Know this material except for a few points. Read the sections about the ones you missed.
13–15	Need to check carefully on the sections you missed.
0–12	Need to work with this chapter to refresh your memory and improve your skills.

6.1   What Is A Fraction?

Mrs. White baked a custard pie and cut it into 8 equal slices. The relationship of *one* of these slices to the whole pie can be represented by a *unit fraction*. The unit fraction in this case is:

$$\frac{1}{8} = \frac{\text{one of the slices}}{\text{the number of slices in the whole pie}}$$

and is illustrated as:

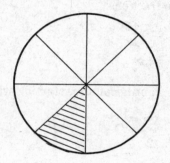

 A fraction can represent the whole pie, $\frac{8}{8}$, or a part of the whole pie, such as $\frac{5}{8}$.

 A fraction can also be one of the parts of a larger quantity or group. What fractional part is 3 men out of a group of 8 men?

$\frac{3}{8}$ is the required fraction.

$$\frac{3}{8} = \frac{\text{three of the men}}{\text{all the men}}$$

Example

If Mrs. White has a family of 5 and they each eat one of the 8 equal pieces of the custard pie, what fraction indicates the portion of the pie that was eaten? What fraction represents the remaining portion?

4

Solution

Since the pie was divided into 8 pieces and 5 of them were eaten, then $\frac{5}{8}$ is the fraction representing the eaten part; $\frac{3}{8}$ is the part that still remains. This is pictured below:

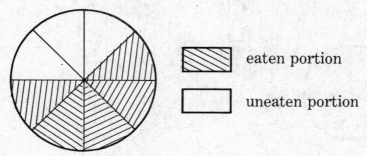

eaten portion

uneaten portion

Thus, a fraction can represent whole parts of a group, 3 men being $\frac{3}{8}$ of the group of 8 men.

 A third meaning of a fraction is revealed by its symbol. Thus, in $\frac{3}{4}$, the line between 3 and 4 means that the 3 is being divided by the 4:

$$3 \div 4 \text{ or } 4\,\overline{)\,3}$$

This is illustrated by dividing 3 units into 4 equal parts:

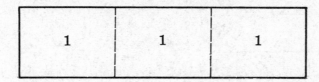

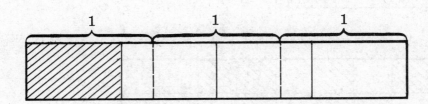

The shaded portion represents $\frac{1}{4}$ of 3 or $\frac{3}{4}$.

 The last meaning of a fraction is a comparison between two numbers. Comparing 3 to 4 is written as $\frac{3}{4}$. This is called a *ratio* and will be studied later in the book.

Practice Exercise 64

For each diagram in the first five problems, answer the following questions:

    a.   Into how many equal parts has the figure been divided?
    b.   Write the unit fraction associated with one of the parts.
    c.   Write the fraction that represents the shaded portion of each figure.

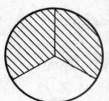

a. _____

b. _____

c. _____

2.

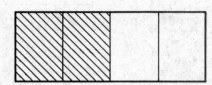

a. _____

b. _____

c. _____

3.

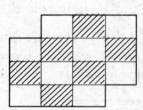

a. _____

b. _____

c. _____

4.

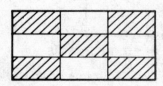

a. _____

b. _____

c. _____

5.

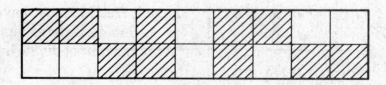

a. _____

b. _____

c. _____

6. Which is the *unit fraction*?

    a. $\dfrac{3}{8}$ b. $\dfrac{2}{5}$ c. $\dfrac{1}{3}$ d. $\dfrac{3}{1}$

_____

7. Mrs. White used 5 eggs to make a cake. Write the fraction that represents the part of a dozen eggs used. (12 eggs = 1 dozen) _____

8. Joe took 7 foul shots in a basketball game. Write the fraction of the total number of shots that were successful if he made 4 baskets. _____

9. The fraction $\frac{5}{8}$ means

    a. $5\overline{)8}$ b. $8\overline{)5}$ _____

10. The ratio 6 ft. to 9 ft. is written as the fraction

    a. $\frac{9}{6}$ b. $\frac{6}{9}$ c. Neither of these   _____

6.2   The Language Of A Fraction

Example

Three of the four cupcakes were eaten at dinner. Write the fraction which represents the portion of the cupcakes which were eaten.

Solution

Since there were originally 4 cupcakes and 3 were eaten, then $\frac{3}{4}$ is the fraction representing the portion that was eaten.

$$\frac{3}{4} = \frac{\text{the number of eaten cupcakes}}{\text{the total number of cupcakes}} = \frac{\text{numerator}}{\text{denominator}}$$

The 4 tells us the total number of parts into which the whole has been divided. This part of a fraction is called the *denominator*. The 3 is called the *numerator;* it tells us the number of parts we are talking about. Both the numerator and the denominator of a fraction are called *terms* of the fraction. The unit fraction must have a numerator equal to 1. A fraction may have any value for the denominator except 0. Remember, in the last chapter in division, it was shown that you cannot divide by 0; therefore no meaningful fraction can have a denominator of 0.

$$\frac{3}{4} = \frac{\text{numerator}}{\text{denominator}}$$

You must know the following terms to understand fractions.

6.3   What Kinds Of Fractions Are There?

There are two kinds of fractions. The fraction that is most common to us is the *proper fraction*. These are all proper fractions: $\frac{3}{4}$, $\frac{5}{10}$, $\frac{7}{8}$. *Proper fractions have a numerator which is less than the denominator*. Thus, the value of any proper fraction must be less than 1.

The second type of fraction has a numerator equal to or greater than the denominator. Examples of *improper fractions* are: $\frac{6}{6}$, $\frac{9}{4}$, $\frac{32}{5}$. *Improper fractions have a value equal to or greater than 1*.

Proper fraction   *Illustration (shaded part)*

Value less than 1

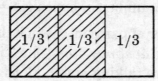

two-thirds
$= \frac{2}{3}$

Improper fraction

a.   Value equal to 1

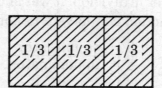

three-thirds
$= \frac{3}{3}$

b. Value greater than 1

 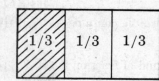 four-thirds $= \frac{4}{3}$

Practice Exercise 65

1. Identify each of these as a *proper* or an *improper* fraction:

   a. $\frac{5}{2}$ _____

   b. $\frac{4}{5}$ _____

   c. $\frac{9}{10}$ _____

   d. $\frac{6}{6}$ _____

   e. $\frac{7}{4}$ _____

In problems 2 through 4 assume that each separate rectangle or triangle represents a unit, 1. The combined shaded pieces in each diagram then represent a certain fractional part of a unit.

   a. Write each fraction.
   b. Identify the kind of fraction it is.

2. 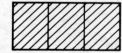 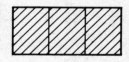 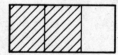   a. _____

   b. _____

3.  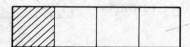   a. _____

   b. _____

4.   a. _____

   b. _____

5. a. An entire pie contains $\frac{?}{8}$.

    a. _____

    b. What kind of fraction is this?

    b. _____

6. A carpenter marked off 7 of the equal divisions on a 12 in. ruler.

    a. What fractional part of the ruler did he mark off?

    a. _____

    b. What kind of fraction do we call this?

    b. _____

6.4   Simplifying Fractions

Which fraction is larger: $\frac{4}{8}$ or $\frac{1}{2}$?

$\frac{4}{8}$ is illustrated as:

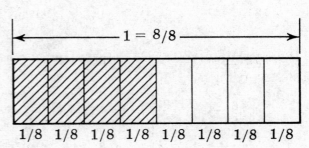

$\frac{4}{8} = \frac{\text{the number of shaded parts}}{\text{the total number of parts}}$

If we assume the whole is the same for both fractions, then

$\frac{1}{2}$ is illustrated as:

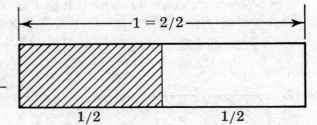

$\frac{1}{2} = \frac{\text{the number of shaded parts}}{\text{the total number of parts}}$

Which fraction is the larger? Obviously, neither one since they are both equal to each other. Thus:

$$\frac{4}{8} = \frac{1}{2}$$

These fractions can be found on a ruler along with other fractions as well. Look at an enlargement of 1 in. on a ruler. Divide it into 2 equal parts. Each part is marked.

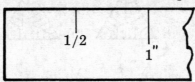

Divide it into 4 equal parts. Each part is $\frac{1}{4}$ of an inch and marked.

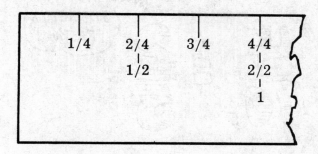

Obviously,

$$\frac{2}{4} = \frac{1}{2} \quad \text{and} \quad \frac{4}{4} = \frac{2}{2} = 1$$

If we continue the division into 8 equal parts, each part is $\frac{1}{8}$ of an inch and is marked accordingly.

Again,

$$\frac{2}{8} = \frac{1}{4}$$

$$\frac{4}{8} = \frac{2}{4} = \frac{1}{2}$$

$$\frac{6}{8} = \frac{3}{4}$$

$$\frac{8}{8} = \frac{4}{4} = \frac{2}{2} = 1$$

Note that in each of the four cases, a fraction with larger terms can be changed to one with smaller terms to which it is equivalent by dividing both terms of the original fraction by the same number, thus reducing its size. The fraction $\frac{1}{2}$ as compared to $\frac{2}{4}$ or $\frac{4}{8}$ is written in its *simplest form*. Some people call it a *reduced fraction*.

11

The fraction $\frac{2}{4}$ is not a fraction in its simplest form, for both the numerator and denominator can be divided by 2. To learn how to simplify a fraction look at this problem:

Problem	Step 1	Step 2
$\frac{2}{4} =$	$\frac{2 \div 2}{4 \div 2} =$	$\frac{1}{2}$
	Find the *largest* number which divides evenly into both the numerator and denominator. In this case, the number is 2.	Divide each term of the fraction by 2. The fraction is equal to $\frac{1}{2}$

Thus: $\frac{2}{4} = \frac{1}{2}$.

No number other than 1 divides evenly into both terms of the fraction $\frac{1}{2}$; therefore; $\frac{1}{2}$ cannot be simplified further. Another fraction written in simplest form is $\frac{8}{15}$.

Example

Simplify the fraction $\frac{3}{6}$.

Solution

Problem	Step 1	Step 2
$\frac{3}{6} =$	$\frac{3 \div 3}{6 \div 3} =$	$\frac{1}{2}$
	3 is the *largest* number which divides evenly (without leaving a remainder) into both the numerator (3) and the denominator (6).	Divide each term by 3: $\frac{3}{6} = \frac{1}{2}$

Thus: $\frac{3}{6} = \frac{1}{2}$.

Practice Exercise 66

Write each fraction in its simplest (or reduced) form. Some fractions cannot be simplified.

1. $\frac{2}{8} =$ 2. $\frac{5}{10} =$ 3. $\frac{4}{20} =$ 4. $\frac{3}{12} =$ 5. $\frac{6}{12} =$

6. $\frac{3}{13} =$ 7. $\frac{8}{16} =$ 8. $\frac{8}{18} =$ 9. $\frac{12}{17} =$ 10. $\frac{28}{32} =$

11. $\frac{6}{15} =$ 12. $\frac{9}{12} =$ 13. $\frac{14}{28} =$ 14. $\frac{10}{15} =$ 15. $\frac{21}{24} =$

Example

Simplify the fraction $\frac{26}{39}$.

Solution

Is there a number which will divide evenly into both 26 and 39? If there is, it is fairly difficult to find. Remember, if there is no number other than 1 which divides into both terms of the fraction, then the fraction cannot be simplified. If there is such a number, then what is it and how do you find it?

The numerator 26 has a number of divisors. Divisors are numbers which will divide evenly into a number. The divisors of 26 are 1, 2, 13, and 26. Each divisor is called a *factor* of the number. Thus, 1, 2, 13, and 26 are all *factors* of 26.

The denominator 39 has factors of 1, 3, 13, and 39. Thus the numbers 26 and 39 have a *common factor* or common divisor of 13. The number 13 can divide evenly into both terms of the fraction, permitting you to simplify the fraction like this:

Problem	Step 1	Step 2
$\dfrac{26}{39} =$	$\dfrac{26 \div 13}{39 \div 13} =$	$\dfrac{2}{3}$
	Each term can be divided evenly by 13.	Dividing each term results in the fraction $\dfrac{2}{3}$

Thus: $\dfrac{26}{39} = \dfrac{2}{3}$.

Every number has a common factor or divisor of 1. Since division by 1 results in the same quotient as the original number, there is no point in dividing by 1.

Example

Simplify the fraction $\dfrac{33}{49}$.

Solution

The numerator 33 has factors of 1, 3, 11, and 33. The denominator 49 has factors of 1, 7, and 49. Since they have no common factors or divisors other than 1, the fraction is already in its simplest form.

Practice Exercise 67

Reduce, if possible, each of the following fractions to its simplest form (lowest term).

1. $\dfrac{25}{50} =$ 2. $\dfrac{44}{77} =$ 3. $\dfrac{35}{47} =$ 4. $\dfrac{45}{63} =$ 5. $\dfrac{35}{49} =$

6. $\dfrac{34}{68} =$ 7. $\dfrac{25}{95} =$ 8. $\dfrac{38}{68} =$ 9. $\dfrac{21}{35} =$ 10. $\dfrac{42}{48} =$

6.5 Equivalent Fractions

When studying the ruler in Section 6.4, you learned that $\dfrac{4}{8} = \dfrac{2}{4} = \dfrac{1}{2}$. Each fraction is written differently but the value of all three fractions is the same. You learned that $\frac{4}{8}$ and $\frac{2}{4}$ can be simplified to $\frac{1}{2}$ by dividing both terms of the fraction by the largest common divisor. Reversing this procedure will allow you to find other fractions with larger terms which are equivalent to a given fraction. This is illustrated below.

Example

Find two fractions equivalent to $\frac{1}{3}$.

Solution

Reversing the procedure means that you must multiply, rather than divide, each term of the fraction by a common number. This common number is called a *common multiplier*. Suppose your common multiplier is 2. This is illustrated as

$$\frac{1 \times 2}{3 \times 2} = \frac{2}{6}$$

Another multiplier can be 4. Thus:

$$\frac{1 \times 4}{3 \times 4} = \frac{4}{12}$$

$$\text{Is } \frac{1}{3} = \frac{2}{6} = \frac{4}{12}?$$

Are these three fractions equivalent fractions? Picture each of these fractions as parts of equal sized rectangles.

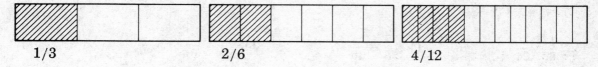

| 1/3 | 2/6 | 4/12 |

The shaded portions of each rectangle are equal despite the fact that the fractions contain different terms. The form of the fraction changes but the value remains the same.

Problem	Step 1	Step 2
$\frac{1}{3} = \frac{?}{6}$	$\frac{1 \times 2}{3 \times 2} = \frac{?}{6}$	$\frac{2}{6}$
	Determine the value of the common multiplier: $3 \times ? = 6; ? = 2$	Multiply both terms of the fraction by 2. Thus: $\frac{1}{3} = \frac{2}{6}$

Thus: $\frac{1}{3} = \frac{2}{6}$.

Problem	Step 1	Step 2
$\frac{3}{4} = \frac{?}{32}$	$\frac{3 \times 8}{4 \times 8} = \frac{?}{32}$	$\frac{24}{32}$
	Determine the value of the common multiplier: $4 \times ? = 32; ? = 8$	Multiply both terms of the fraction by 8.

Thus: $\frac{3}{4} = \frac{24}{32}$.

Practice Exercise 68

Change each of the following fractions to the new denominator. (Change each fraction to higher terms.)

1. $\frac{2}{3} = \frac{}{6}$

2. $\frac{2}{5} = \frac{}{35}$

3. $\frac{3}{5} = \frac{}{25}$

4. $\frac{3}{4} = \frac{}{8}$

5. $\frac{1}{8} = \frac{}{24}$

6. $\frac{1}{2} = \frac{}{8}$

7. $\frac{2}{9} = \frac{}{36}$

8. $\frac{3}{8} = \frac{}{40}$

9. $\frac{9}{10} = \frac{}{40}$

10. $\frac{5}{6} = \frac{}{18}$

11. $\frac{2}{5} = \frac{}{15}$

12. $\frac{2}{3} = \frac{}{12}$

13. $\frac{1}{6} = \frac{}{24}$

14. $\frac{7}{8} = \frac{}{16}$

15. $\frac{3}{4} = \frac{}{16}$

Terms You Should Remember

Divisor   The quantity by which the dividend is to be divided.

Factor   One of two or more quantities having a designated product: 2 and 3 are factors of 6.

Simplest form   The way of writing a fraction that uses the smallest numbers (lowest terms).

Equivalent   Equal in value.

> *Lowest terms*   The parts of a fraction whose terms cannot be reduced because they have no common factor.
>
> *Higher terms*   The parts of a fraction whose terms can be reduced because they have a common factor other than 1.

6.6   Which Fraction Is Larger Or Smaller?

Which would you rather have, $\frac{1}{4}$ of a dollar or $\frac{5}{12}$ of a dollar?

Certainly, you would choose the larger fractional part but which fraction is the larger?

If the denominators of both fractions are the same, it is an easy task to find the larger or smaller fraction. Look at the example below:

$$\frac{5}{12} = \frac{5 \text{ of those parts}}{\text{number the whole is divided into}}$$

$$\frac{3}{12} = \frac{3 \text{ of those parts}}{\text{number the whole is divided into}}$$

If the whole in both fractions is the same, then obviously $\frac{5}{12}$ is greater than $\frac{3}{12}$.

Recall the inequality symbol:

$$\frac{5}{12} > \frac{3}{12}$$

What happens if the denominations are not alike? Which fraction is larger, $\frac{1}{4}$ or $\frac{5}{12}$?

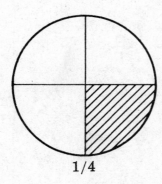

1/4

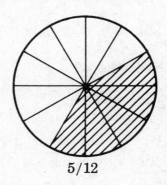

5/12

You can see that
$$\frac{5}{12} > \frac{1}{4}$$

This picture or illustration method is not satisfactory when the two fractions are very close in value.

Another method you can use to compare fractions is:

 Change each fraction to the same denominator.

 Choose as the larger fraction the one which has the larger numerator.

Sometimes when changing each fraction to the same denominator you change the denominator with the smaller number to the same number as the one with the larger denominator. Look at the next example:

Problem	Step 1	Step 2	Step 3
Which is larger? $\frac{5}{12}$ or $\frac{1}{4}$	$\frac{1}{4} = \frac{?}{12}$	$\frac{1 \times 3}{4 \times 3} = \frac{3}{12}$	$\frac{5}{12} > \frac{3}{12}$ or $\frac{5}{12} > \frac{1}{4}$
	Change the fraction with the smaller denominator, $\frac{1}{4}$, to an equivalent fraction whose denominator is 12ths.	Multiply both terms of the fraction by 3. Thus: $\frac{1}{4} = \frac{3}{12}$	Compare the fractions: $\frac{5}{12} > \frac{3}{12}$ or $\frac{5}{12} > \frac{1}{4}$

Thus: $\frac{5}{12} > \frac{1}{4}$.

Practice Exercise 69

Which is the larger fraction?

1. $\frac{7}{12}$ or $\frac{9}{12}$ _____

2. $\frac{5}{6}$ or $\frac{4}{6}$ _____

3. $\frac{2}{3}$ or $\frac{5}{12}$ _____

4. $\frac{3}{5}$ or $\frac{11}{20}$ _____

Which is the smaller fraction?

5. $\dfrac{9}{10}$ or $\dfrac{8}{10}$ _____

6. $\dfrac{13}{16}$ or $\dfrac{11}{16}$ _____

7. $\dfrac{2}{3}$ or $\dfrac{5}{24}$ _____

8. $\dfrac{5}{6}$ or $\dfrac{7}{18}$ _____

Example

Compare the fractions $\dfrac{1}{2}$ and $\dfrac{1}{3}$.

Solution

Sometimes it is very difficult to change the denominator of the fraction with the smaller number to the same number as the one with the larger denominator. In the case of $\dfrac{1}{2} = \dfrac{?}{3}$, or $2 \times ? = 3$, there is no whole number which can be used, so you must find another way of doing this example. You must find a *common denominator* for both fractions. A *common denominator* is a number that is divisible by both denominators. In this example, 6 is divisible by both 2 and 3. Other numbers are also divisible by 2 and 3, but 6 is the *lowest (or least) common denominator* (abbreviated as L.C.D.).

Problem	Step 1	Step 2	Step 3
Compare the fractions:	$\dfrac{1}{2} = \dfrac{?}{6}$	$\dfrac{1 \times 3}{2 \times 3} = \dfrac{3}{6}$	$\dfrac{3}{6} > \dfrac{2}{6}$
$\dfrac{1}{2}$ and $\dfrac{1}{3}$	$\dfrac{1}{3} = \dfrac{?}{6}$	$\dfrac{1 \times 2}{3 \times 2} = \dfrac{2}{6}$	or $\dfrac{1}{2} > \dfrac{1}{3}$

Thus: $\dfrac{1}{2} > \dfrac{1}{3}$.

Example

Arrange the fractions $\dfrac{1}{2}$, $\dfrac{5}{8}$, $\dfrac{2}{3}$ in order of size, the largest first.

Solution

First find the *lowest common denominator* for the three fractions. What number will 2, 8, and 3 divide into evenly?

The fraction $\dfrac{1}{2}$ can be changed to an equivalent fraction whose denominator is 2, 4, 6, 8, 10, 12, 14, 16, 18, 20, 22, (24), 26, etc.

The fraction $\frac{5}{8}$ can be changed to an equivalent fraction whose denominator is 8, 16, (24), 32, 40, etc.

The fraction $\frac{2}{3}$ can be changed to an equivalent fraction whose denominator is 3, 6, 9, 12, 15, 18, 21, (24), 27, 30, etc.

The numbers 2, 8, and 3 will all divide evenly into 24. Thus, 24 is the *lowest common denominator*.

Problem	Step 1	Step 2	Step 3
Compare: $\frac{1}{2}$, $\frac{5}{8}$, $\frac{2}{3}$	$\frac{1}{2} = \frac{?}{24}$ $\frac{5}{8} = \frac{?}{24}$ $\frac{2}{3} = \frac{?}{24}$	$\frac{1 \times 12}{2 \times 12} = \frac{12}{24}$ $\frac{5 \times 3}{8 \times 3} = \frac{15}{24}$ $\frac{2 \times 8}{3 \times 8} = \frac{16}{24}$	$\frac{16}{24} > \frac{15}{24} > \frac{12}{24}$ or $\frac{2}{3} > \frac{5}{8} > \frac{1}{2}$

Thus: $\frac{2}{3} > \frac{5}{8} > \frac{1}{2}$.

Practice Exercise 70

Find the lowest common denominator for each group of examples.

1. $\frac{2}{3}$ and $\frac{5}{9}$ _____

2. $\frac{1}{4}$ and $\frac{2}{3}$ _____

3. $\frac{1}{2}$ and $\frac{4}{5}$ _____

4. $\frac{4}{5}$ and $\frac{2}{3}$ _____

5. $\frac{3}{10}$ and $\frac{7}{30}$ _____

6. $\frac{4}{5}$ and $\frac{4}{15}$ _____

Arrange the fractions in order of size, the smallest one first.

7. $\frac{1}{2}$, $\frac{5}{6}$, $\frac{3}{5}$ _____

8. $\frac{4}{5}$, $\frac{5}{12}$, $\frac{5}{6}$ _____

Arrange the fractions in order of size, the largest one first.

9. $\frac{1}{2}, \frac{4}{5}, \frac{2}{3}$ _____

10. $\frac{4}{15}, \frac{3}{10}, \frac{7}{30}$ _____

6.7 Changing Improper Fractions To Mixed Numbers

You learned earlier that the *value of an improper fraction must be equal to or greater than 1*. Fractions such as $\frac{4}{4}$, $\frac{5}{4}$, $\frac{6}{4}$, etc., are *improper fractions*.

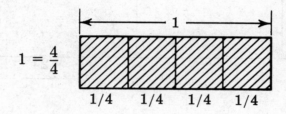

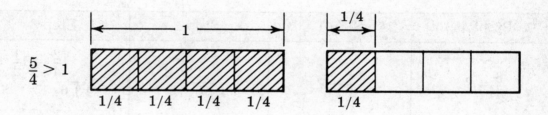

$\frac{5}{4} = 1$ and $\frac{1}{4}$, which is written as $1\frac{1}{4}$. This is called a *mixed number*. It is made up of two parts:

$$1 = \text{whole number}$$
$$\frac{1}{4} = \text{proper fraction}$$

An improper fraction can be changed into a *mixed number* either by using the method just shown or by dividing. Recall that the line separating the 5 and 4 means that the numerator (5) is being divided by the denominator (4):

$$5 \div 4 \quad \text{or} \quad 4\overline{)5}$$

Problem	Step 1	Step 2	Step 3
$\frac{5}{4} = ?$	$4\overline{)5}$	$\overset{1\ R\ 1}{4\overline{)5}}$	$\overset{1\frac{1}{4}}{4\overline{)5}}$
	Write the improper fraction as a division example: $4\overline{)5}$	Divide the denominator into the numerator: $\overset{1\ R\ 1}{4\overline{)5}}$	Write the remainder as the numerator and the divisor as the denominator of a fraction $\frac{1}{4}$ *Write it in its simplest form.*

Thus: $\frac{5}{4} = 1\frac{1}{4}$.

Problem	Step 1	Step 2	Step 3
$\frac{6}{4} = ?$	$4\overline{)6}$	$\overset{1\ R\ 2}{4\overline{)6}}$	$\overset{1\frac{2}{4}\ =\ 1\frac{1}{2}}{4\overline{)6}}$
	Divide the denominator (4) into the numerator (6): $4\overline{)6}$	Divide: 1 R 2	Write the remainder as the numerator and the divisor as the denominator of a fraction *in its simplest form:* $\frac{2}{4} = \frac{1}{2}$

Thus: $\frac{6}{4} = 1\frac{1}{2}$.

Practice Exercise 71

Write as a simple fraction, the shaded portion of the following, assuming that each circle, rectangle, or triangle represents a whole unit.

1.   _____

2. 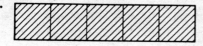 _____

3. _____

Write each of these mixed numbers in its simplest form.

4. $1\frac{6}{8} =$ 5. $2\frac{5}{15} =$ 6. $5\frac{2}{6} =$

Write each of these improper fractions as a whole or mixed number. Each fraction should be written in simplest form.

7. $\frac{13}{5} =$ 8. $\frac{2}{2} =$ 9. $\frac{13}{2} =$ 10. $\frac{8}{6} =$

11. $\frac{41}{8} =$ 12. $\frac{32}{3} =$ 13. $\frac{19}{4} =$ 14. $\frac{5}{3} =$

15. $\frac{17}{8} =$ 16. $\frac{9}{3} =$ 17. $\frac{80}{9} =$ 18. $\frac{37}{5} =$

19. $\frac{43}{2} =$ 20. $\frac{32}{5} =$ 21. $\frac{18}{4} =$ 22. $\frac{5}{5} =$

23. $\frac{19}{3} =$ 24. $\frac{12}{4} =$ 25. $\frac{22}{6} =$

23

6.8 Changing Mixed Numbers To Improper Fractions

A mixed number consists of a whole number and a proper fraction. A mixed number can be changed into an improper fraction only. The mixed number $2\frac{1}{2}$ is pictured as:

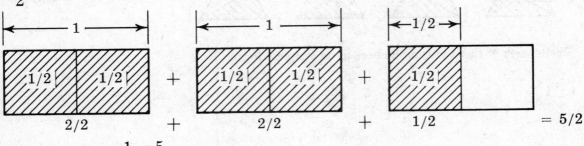

Therefore: $2\frac{1}{2} = \frac{5}{2}$.

A mixed number can be changed into an improper fraction by adding the whole number part to the fraction part after first writing the whole number part as a fraction. Recall that a whole number may be written as a fraction with a denominator equal to 1.

Problem	Step 1	Step 2	Step 3
$2\frac{1}{2} = \frac{?}{2}$	$\frac{2 \times 2}{1 \times 2} = \frac{4}{2}$	$\frac{4}{2} + \frac{1}{2} = \frac{5}{2}$	$2\frac{1}{2} = \frac{5}{2}$
	Multiply the terms of the fraction $\frac{2}{1}$ by 2: $\frac{2 \times 2}{1 \times 2} = \frac{4}{2}$	Add the fraction $\frac{1}{2}$ to the fraction $\frac{4}{2}$: $\frac{4}{2} + \frac{1}{2} = \frac{5}{2}$	Thus: $2\frac{1}{2} = \frac{5}{2}$

Thus: $2 = \frac{2}{1}$.

Example

Change $4\frac{2}{3}$ into an improper fraction.

Solution

Problem	Step 1	Step 2	Step 3
$4\frac{2}{3} = \frac{?}{3}$	$\frac{4 \times 3}{1 \times 3} = \frac{12}{3}$	$\frac{12}{3} + \frac{2}{3} = \frac{14}{3}$	$4\frac{2}{3} = \frac{14}{3}$
	Multiply the terms of the fraction $\frac{4}{1}$ by 3: $\frac{4 \times 3}{1 \times 3} = \frac{12}{3}$	Add the fraction $\frac{2}{3}$ to the fraction $\frac{12}{3}$: $\frac{12}{3} + \frac{2}{3} = \frac{14}{3}$	Thus: $4\frac{2}{3} = \frac{14}{3}$

Thus: $4\frac{2}{3} = \frac{14}{3}$.

Practice Exercise 72

Change each of the following whole or mixed numbers into improper fractions.

1. $4\frac{1}{9} = \frac{}{9}$

2. $3\frac{1}{2} = \frac{}{2}$

3. $6\frac{3}{5} = \frac{}{5}$

4. $6 = \frac{}{5}$

5. $8\frac{3}{4} = \frac{}{4}$

6. $10\frac{3}{7} = \frac{}{7}$

7. $7\frac{1}{10} = \frac{}{10}$

8. $11\frac{3}{5} = \frac{}{5}$

9. $12\frac{5}{8} = \frac{}{8}$

10. $4\frac{5}{7} = \frac{}{7}$

11. $6\frac{1}{6} = \frac{}{6}$

12. $5 = \frac{}{4}$

13. $8 = \frac{}{3}$

14. $14\frac{2}{3} = \frac{}{3}$

15. $23\frac{13}{16} = \frac{}{16}$

Review Of Important Ideas

Some of the most important ideas in Chapter 6 were:

 A fraction is a quotient of two quantities.

 There are proper and improper fractions.

 Fractions can be *reduced*, or written in simplest form.

 Equivalent fractions are fractions whose values are the same but whose respective numerators and denominators may be different.

Check What You Have Learned

The following test lets you see how well you have learned the ideas in Chapter 6. Before you begin, though, there is one hint you should be aware of.

You see how the method of doing an example depends on the basic *how and why* rather than on memorization. Nevertheless, there are times when memorizing is important. Memorizing the basic number facts of addition, subtraction, multiplication, and division is a necessity. If you attempt to memorize methods for solving each example or problem, you are doomed to fail. You will always be dependent upon the memorized fact and, if you forget that, you will be unable to solve the problem. Don't let this happen to you. Know the *how and why*, and then the solution to each problem becomes meaningful and understandable.

Posttest 6

Have you worked out your difficulties with the simple operations of fractions? Try to get a satisfactory grade on this posttest.

Reduce all fractions to their simplest form.

1. Using the diagram shown, write as a fraction the portion of the rectangle that has been shaded. _____

2. A shopper buys a package of 8 apples and eats 6. What fractional part has been eaten? _____

3. In the fraction $\frac{15}{16}$, name the numerator and the denominator.

    15 is the _____

    16 is the _____

4. Identify each fraction below as a *proper* or an *improper* fraction.

    a. $\frac{2}{3}$

    a. _____

    b. $\frac{5}{5}$

    b. _____

    c. $\frac{6}{5}$

    c. _____

5. A line is divided into 3 equal segments. Using a fraction, tell what name is given to one of these parts. _____

6. If there are 60 minutes in each hour, what fractional part of an hour is 40 minutes? _____

7. At a party of 25 adults, 15 are women.

    a. What fraction of the party is women?   a. _____

    b. What fraction of the party is men?   b. _____

8. Write the fraction $\frac{24}{32}$ in its simplest form.

                                                      _____

9. To change $\frac{3}{5}$ to an equivalent fraction with a new denominator of
    25, you must multiply the numerator and denominator by a cer-
    tain number. What is that number?   _____

10. Change the improper fraction $\frac{23}{8}$ to a mixed number.

                                                      _____

11. Change the mixed number $4\frac{2}{7}$ to an improper fraction.

                                                      _____

12. In a mathematics class there are 2 women with blonde hair out
    of a total of 8 women in the class.

    a. What fraction of the class has blonde hair?   a. _____

    b. What fraction of the class does not have blonde hair?   b. _____

13. Find the lowest common denominator for the fractions:
    $$\frac{3}{8}, \frac{1}{3}, \text{ and } \frac{5}{12}$$
                                                      _____

14. Which is the larger of the two fractions:
    $$\frac{3}{4} \text{ or } \frac{19}{24}?$$
                                                      _____

15. Arrange these three fractions in order of size, starting with the
    smallest one:
    $$\frac{7}{8}, \frac{13}{16}, \frac{3}{4}$$
                                                      _____

28

1. $\frac{5}{7}$

2. $\frac{6}{8} = \frac{3}{4}$

3. 15 is the numerator
   16 is the denominator

4. a. proper
   b. improper
   c. improper

5. $\frac{1}{3}$

6. $\frac{40}{60} = \frac{40 \div 20}{60 \div 20} = \frac{2}{3}$

7. a. $\frac{15}{25} = \frac{3}{5}$ women

   b. $\begin{array}{r} 25 \\ -15 \\ \hline 10 \end{array}$ $\quad \frac{10}{25} = \frac{2}{5}$ men

8. $\frac{24 \div 8}{32 \div 8} = \frac{3}{4}$

9. 5

10. $\frac{23}{8} = 8\overline{)23}^{\;2\,R\,7\;=\;2\frac{7}{8}}$

11. $4\frac{2}{7} = \frac{4 \times 7}{1 \times 7} + \frac{2}{7} = \frac{28}{7} + \frac{2}{7} = \frac{30}{7}$

12. a. $\frac{2 \div 2}{8 \div 2} = \frac{1}{4}$

    b. $\begin{array}{r} 8 \\ -2 \\ \hline 6 \end{array}$ $\quad \frac{6 \div 2}{8 \div 2} = \frac{3}{4}$

13. L.C.D. = 24

14. $\frac{3 \times 6}{4 \times 6} = \frac{18}{24}$

    $\frac{19}{24} > \frac{18}{24} \qquad \frac{19}{24} > \frac{3}{4}$

15. $\frac{7 \times 2}{8 \times 2} = \frac{14}{16}$

    $\frac{3 \times 4}{4 \times 4} = \frac{12}{16}$

    Thus:

    $\frac{12}{16} < \frac{13}{16} < \frac{14}{16} \quad$ or

    $\frac{3}{4} < \frac{13}{16} < \frac{7}{8}$

In counting up your answers, remember that there were 20 separate answers.

A Score of	Means That You
18–20	Did very well. You can move to Chapter 7.
16–17	Know this material except for a few point. Reread the sections about the ones you missed.
13–15	Need to check carefully on the sections you missed.
0–12	Need to review the chapter to refresh your memory and improve your skills.

Questions	Are Covered in Section
1–2, 5–7, 12	6.1
3	6.2
4	6.3
8	6.4

ANSWERS FOR CHAPTER 6

PRETEST 6

1. $\frac{4}{6} = \frac{2}{3}$

2. $\frac{2}{8} = \frac{1}{4}$

3. 3 is the numerator
   16 is the denominator

4. a. improper
   b. proper
   c. improper

5. $\frac{1}{12}$

6. $\begin{array}{r} 5 \\ -2 \\ \hline 3 \end{array}$ $\quad \frac{3}{5}$

7. a. $\frac{17}{27}$ men

   b. $\begin{array}{r} 27 \\ -17 \\ \hline 10 \end{array}$ $\quad \frac{10}{27}$ women

8. $\frac{3}{8}$

9. 4

10. $2\frac{4}{5}$

11. $\frac{10}{3}$

12. a. $\frac{10 \div 2}{24 \div 2} = \frac{5}{12}$

    b. $\begin{array}{r} 24 \\ -10 \\ \hline 14 \end{array}$ $\quad \frac{14 \div 2}{24 \div 2} = \frac{7}{12}$

13. L.C.D. = 12

14. $\frac{11}{12} > \frac{7}{8}$

15. $\frac{5}{12}, \frac{1}{2}, \frac{2}{3}$

PRACTICE EXERCISE 64

1. a. 3
   b. $\frac{1}{3}$
   c. $\frac{2}{3}$

2. a. 4
   b. $\frac{1}{4}$
   c. $\frac{2}{4}$

3. a. 14
   b. $\frac{1}{14}$
   c. $\frac{6}{14}$

4. a. 9

    b. $\frac{1}{9}$

    c. $\frac{5}{9}$

5. a. 18

    b. $\frac{1}{18}$

    c. $\frac{10}{18}$

6. (c) $\frac{1}{3}$

7. $\frac{5}{12}$

8. $\frac{4}{7}$

9. (b) $8\overline{)5}$

10. (b) $\frac{6}{9}$

PRACTICE EXERCISE 65

1. a. improper
    b. proper
    c. proper
    d. improper
    e. improper

2. a. $\frac{5}{4}$
    b. improper

3. a. $\frac{8}{3}$
    b. improper

4. a. $\frac{12}{3}$
    b. improper

5. a. 8 ($\frac{8}{8}$ = entire pie)
    b. improper

6. a. $\frac{7}{12}$
    b. proper

PRACTICE EXERCISE 66

1. $\frac{1}{4}$ 2. $\frac{1}{2}$ 3. $\frac{1}{5}$ 4. $\frac{1}{4}$ 5. $\frac{1}{2}$

6. $\frac{3}{13}$ 7. $\frac{1}{2}$ 8. $\frac{4}{9}$ 9. $\frac{12}{17}$ 10. $\frac{7}{8}$

11. $\frac{2}{5}$ 12. $\frac{3}{4}$ 13. $\frac{1}{2}$ 14. $\frac{2}{3}$ 15. $\frac{7}{8}$

PRACTICE EXERCISE 67

1. $\frac{1}{2}$ 2. $\frac{4}{7}$ 3. $\frac{35}{47}$ 4. $\frac{5}{7}$ 5. $\frac{5}{7}$

6. $\frac{1}{2}$ 7. $\frac{5}{19}$ 8. $\frac{19}{34}$ 9. $\frac{3}{5}$ 10. $\frac{7}{8}$

PRACTICE EXERCISE 68

1. $\frac{4}{6}$ 2. $\frac{14}{35}$ 3. $\frac{15}{25}$ 4. $\frac{6}{8}$ 5. $\frac{3}{24}$

6. $\frac{4}{8}$ 7. $\frac{8}{36}$ 8. $\frac{15}{40}$ 9. $\frac{36}{40}$ 10. $\frac{15}{18}$

11. $\frac{6}{15}$ 12. $\frac{8}{12}$ 13. $\frac{4}{24}$ 14. $\frac{14}{16}$ 15. $\frac{12}{16}$

PRACTICE EXERCISE 69

1. $\frac{9}{12}$
2. $\frac{5}{6}$
3. $\frac{2}{3}$
4. $\frac{3}{5}$
5. $\frac{8}{10}$

6. $\frac{11}{16}$
7. $\frac{5}{24}$
8. $\frac{7}{18}$

PRACTICE EXERCISE 70

1. 9
2. 12
3. 10
4. 15
5. 30

6. 15
7. $\frac{1}{2}, \frac{3}{5}, \frac{5}{6}$
8. $\frac{5}{12}, \frac{4}{5}, \frac{5}{6}$
9. $\frac{4}{5}, \frac{2}{3}, \frac{1}{2}$
10. $\frac{3}{10}, \frac{4}{15}, \frac{7}{30}$

PRACTICE EXERCISE 71

1. $\frac{11}{5}$
2. $\frac{8}{5}$
3. $\frac{11}{3}$
4. $1\frac{3}{4}$
5. $2\frac{1}{3}$

6. $5\frac{1}{3}$
7. $2\frac{3}{5}$
8. 1
9. $6\frac{1}{2}$
10. $1\frac{1}{3}$

11. $5\frac{1}{8}$
12. $10\frac{2}{3}$
13. $4\frac{3}{4}$
14. $1\frac{2}{3}$
15. $2\frac{1}{8}$

16. 3
17. $8\frac{8}{9}$
18. $7\frac{2}{5}$
19. $21\frac{1}{2}$
20. $6\frac{2}{5}$

21. $4\frac{1}{2}$
22. 1
23. $6\frac{1}{3}$
24. 3
25. $3\frac{2}{3}$

PRACTICE EXERCISE 72

1. $\frac{37}{9}$
2. $\frac{7}{2}$
3. $\frac{33}{5}$
4. $\frac{30}{5}$
5. $\frac{35}{4}$

6. $\frac{73}{7}$
7. $\frac{71}{10}$
8. $\frac{58}{5}$
9. $\frac{101}{8}$
10. $\frac{33}{7}$

11. $\frac{37}{6}$
12. $\frac{20}{4}$
13. $\frac{24}{3}$
14. $\frac{44}{3}$
15. $\frac{381}{16}$

You Are On Your Way

You have completed the entire work on whole numbers and you have just entered into the world of fractions. You know that in your everyday life, fractions appear just as frequently as whole numbers. The four operations of addition, subtraction, multiplication, and division of fractions will be studied in the next three chapters.

In the last chapter you learned about fractions and their relationship to whole and mixed numbers. This chapter discusses how to *use* fractions or how to *operate* with them.

The topic is *addition* of fractions, the first operation dealing with fractions. You know you can figure the total of two or more fractions by adding them together. No matter what fractions or how many fractions you are asked to add together, it can be done by following certain basic steps. Check your skill on this topic in the pretest which follows.

See What You Know And Remember — Pretest 7

Work these exercises carefully, doing as many of these problems as you can. Write your answers in simplified form below each question.

1. $\dfrac{3}{8}$
    $+\dfrac{4}{8}$

2. $\dfrac{1}{6}$
    $+\dfrac{1}{6}$

3. $\dfrac{2}{5}$
    $+\dfrac{3}{5}$

4. $\dfrac{7}{9}$
    $+\dfrac{4}{9}$

5. $\dfrac{3}{10}$

$+\dfrac{9}{10}$

6. $4\dfrac{2}{3}$

$+3$

7. 7

$+4\dfrac{1}{2}$

8. $6\dfrac{1}{3}$

$+2\dfrac{1}{3}$

9. $5\dfrac{1}{8}$

$+4\dfrac{3}{8}$

10. $14\dfrac{5}{6}$

$+\ 5\dfrac{1}{6}$

11. $9\dfrac{4}{7}$

$+2\dfrac{5}{7}$

12. $7\dfrac{15}{16}$

$+\ \dfrac{9}{16}$

13. A recipe calls for adding $\dfrac{3}{8}$ of a cup of water now and $\dfrac{1}{8}$ of a cup later. How much water is used in this recipe?

14. Mrs. Pink shops for food and returns with $3\dfrac{1}{4}$ lb. of potatoes, 1 gal. of milk which weighs $3\dfrac{3}{4}$ lb., and a six-pack of soda which weighs $5\dfrac{1}{4}$ lbs. What was the total number of pounds Mrs. Pink carried home?

15. Find the least common denominator for these *three* fractions: $\dfrac{1}{3}$, $\dfrac{3}{4}$, $\dfrac{5}{6}$

16. $\dfrac{1}{2}$

$+\dfrac{3}{8}$

17. $\dfrac{2}{3}$

$+\dfrac{5}{24}$

18. $\dfrac{3}{4}$

$+\dfrac{1}{2}$

19. $\dfrac{1}{2}$

$\dfrac{2}{3}$

$+\dfrac{3}{5}$

20. $3\dfrac{2}{3}$

$+2\dfrac{5}{6}$

21. $7\dfrac{5}{8}$

$5\dfrac{9}{16}$

$+3\dfrac{1}{4}$

22. How many feet of fencing are required to enclose the vegetable garden pictured below?

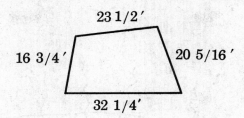

23. How many yards of rope were removed from a spool if $4\frac{5}{6}$ yd. were used on Monday, $10\frac{1}{2}$ yd. on Tuesday, and $7\frac{1}{12}$ yd. on Wednesday?

Check your answers by turning to the end of the chapter. Add up all that you had correct.

A Score of	Means That You
21–23	Did very well. You can move to Chapter 8.
18–20	Know the material except for a few points. Read the sections about the ones you missed.
15–17	Need to check carefully on the sections you missed.
0–14	Need to work with the chapter to refresh your memory and improve your skills.

Questions	Are Covered in Section
1	7.1
2	7.2
3	7.3
4–5	7.4
6–8	7.5
9	7.6
10	7.7
11–12	7.8
13–14	7.9
15–19	7.10
20–21	7.11
22–23	7.12

7.1 Fractions Can Be Added

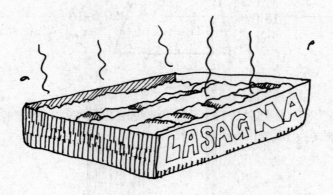

Example

Mrs. Calucci baked a dish of lasagna for her family and cut it into 6 equal portions. If Mrs. Calucci serves 3 portions, or $\frac{3}{6}$, for lunch and 2 portions, or $\frac{2}{6}$, for dinner, what part of the lasagna has she served?

Solution

To find out how much was served, you must find the *sum* of $\frac{3}{6}$ and $\frac{2}{6}$. This means you must *add* the two fractions together. The diagram below will help you to discover how fractions are added.

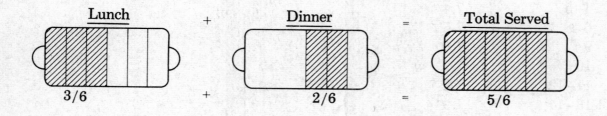

Lunch	+	Dinner	=	Total Served
3/6	+	2/6	=	5/6

Mrs. Calucci served $\frac{5}{6}$ of her dish of lasagna. These fractions are called *like fractions* since they have the same number in the denominator, namely, sixths or 6ths.

To add like fractions, *add* only the numerators; the denominator remains the same. Look at this example:

Problem	Step 1	Step 2
$\dfrac{3}{6} + \dfrac{2}{6} =$	$\dfrac{3+2}{6}$	$\dfrac{5}{6}$
	Add the numerators: $3 + 2 = 5$	Place this sum (5) over the same denominator (6): $\dfrac{5}{6}$

Thus: $\dfrac{3}{6} + \dfrac{2}{6} = \dfrac{5}{6}$.

Terms You Should Remember

Add   To join to another.

Sum   The result of adding two or more numbers.

Like fractions   Fractions whose denominators are the same number.

Numerator   The number in the top portion of a fraction (dividend).

Denominator   The number in the bottom portion of a fraction (divisor).

Practice Exercise 73

Find the *sum* of each of the following problems. Follow the steps shown in the previous example.

1. $\dfrac{1}{5} + \dfrac{2}{5} = \dfrac{}{5}$

2. $\dfrac{1}{3} + \dfrac{1}{3} = \dfrac{}{3}$

3. $\dfrac{2}{4} + \dfrac{1}{4} = \dfrac{}{4}$

4. $\dfrac{4}{7} + \dfrac{1}{7} = \dfrac{}{7}$

5. $\dfrac{2}{9} + \dfrac{4}{9} + \dfrac{1}{9} =$

6. $\dfrac{6}{11} + \dfrac{3}{11} =$

7. $\dfrac{1}{6} + \dfrac{1}{6} + \dfrac{3}{6} =$

8. $\dfrac{1}{8} + \dfrac{2}{8} + \dfrac{4}{8} =$

9. $\frac{5}{12} + \frac{3}{12} + \frac{3}{12} =$ **10.** $\frac{2}{10} + \frac{4}{10} + \frac{1}{10} =$

7.2 After Adding, What Next?

Example

A recipe calls for $\frac{1}{4}$ cup of water now and another $\frac{1}{4}$ cup of water later. How much water is needed in the recipe?

Solution

You must add the fractions $\frac{1}{4} + \frac{1}{4}$ to find how much water is needed. Look at the following example:

Problem	Step 1	Step 2	Step 3
$\frac{1}{4} + \frac{1}{4}$	$\frac{1+1}{4}$	$\frac{2}{4}$	$\frac{2 \div 2}{4 \div 2} = \frac{1}{2}$
	Add the numerators together: $1 + 1 = 2$	Place this sum over the same denominator (4): $\frac{2}{4}$	Write the fraction $\frac{2}{4}$ in its simplest form. Divide the numerator and denominator by 2: $\frac{2}{4} = \frac{1}{2}$

Thus: $\frac{1}{4} + \frac{1}{4} = \frac{1}{2}$.

Practice Exercise 74

Add the fractions and write each sum in its simplest form.

1. $\frac{1}{9} + \frac{2}{9} =$ **2.** $\frac{1}{6} + \frac{3}{6} =$

3. $\frac{3}{8} + \frac{1}{8} =$ **4.** $\frac{3}{10} + \frac{2}{10} =$

5. $\frac{1}{12} + \frac{3}{12} =$

6. $\frac{1}{6} + \frac{1}{6} + \frac{1}{6} =$

7. $\frac{1}{10} + \frac{2}{10} + \frac{5}{10} =$

8. $\frac{7}{12} + \frac{1}{12} + \frac{1}{12} =$

9. $\frac{2}{8} + \frac{3}{8} + \frac{1}{8} =$

10. $\frac{2}{9} + \frac{2}{9} + \frac{2}{9} =$

7.3   Fractions Whose Sum Is One

Example

David jogged $\frac{3}{4}$ of a mile before he got tired and had to rest. He then ran another $\frac{1}{4}$ of a mile. How many miles did he run altogether?

Solution

You must find the sum of the two fractions $\frac{3}{4} + \frac{1}{4}$ to determine the number of miles he ran altogether.

Problem	Step 1	Step 2	Step 3
$\frac{3}{4} + \frac{1}{4}$	$\frac{3 + 1}{4}$	$\frac{4}{4}$	$\frac{4}{4} = 1$
	Add the numerators: $3 + 1 = 4$	Place the sum (4) over the same denominator (4): $\frac{4}{4}$	Simplify the fraction: $\frac{4}{4} = 1$ (Recall that the value of a fraction is 1 if it has the same numerator and denominator. $\frac{0}{0}$ is the exception to this rule.)

Thus: $\frac{3}{4} + \frac{1}{4} = 1$.

Add each of the following fractions. Simplify each sum when it is possible.

1. $\frac{2}{5} + \frac{3}{5} =$

2. $\frac{1}{2} + \frac{1}{2} =$

3. $\frac{3}{8} + \frac{4}{8} =$

4. $\frac{1}{3} + \frac{2}{3} =$

5. $\frac{1}{6} + \frac{5}{6} =$

6. $\frac{1}{7} + \frac{4}{7} + \frac{2}{7} =$

7. $\frac{4}{8} + \frac{3}{8} + \frac{1}{8} =$

8. $\frac{4}{10} + \frac{3}{10} + \frac{2}{10} =$

9. $\frac{4}{12} + \frac{4}{12} + \frac{4}{12} =$

10. $\frac{3}{6} + \frac{1}{6} + \frac{2}{6} =$

7.4   Fractions Whose Sum Is Greater Than One

Example

Mr. Garcia is a carpenter who made two measurements ($\frac{7}{8}$ in. and $\frac{5}{8}$ in.), as shown below on a ruler. What was the total number of inches he measured?

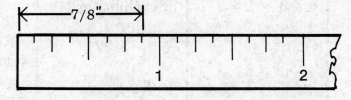

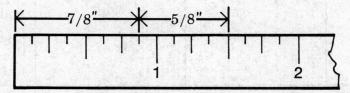

Solution

You must add the fractions $\frac{7}{8}$ and $\frac{5}{8}$ to find the total number of inches.

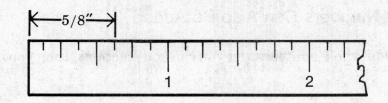

Problem	Step 1	Step 2	Step 3
$\dfrac{7}{8} + \dfrac{5}{8}$	$\dfrac{7+5}{8}$	$\dfrac{12}{8}$	$\begin{array}{r} 1 \text{ R } 4 \\ 8\overline{)12} \end{array}$ or $1\dfrac{4}{8} = 1\dfrac{1}{2}$
	Add the numerators: $7 + 5 = 12$	Place this sum (12) over the same denominator (8): $\dfrac{12}{8}$	Change the improper fraction into a mixed number: $\dfrac{12}{8} = 1\dfrac{4}{8} = 1\dfrac{1}{2}$

Thus: $\dfrac{7}{8} + \dfrac{5}{8} = 1\dfrac{1}{2}$.

Practice Exercise 76

Add each of the following problems. Write each sum in its simplest form.

1. $\dfrac{3}{4} + \dfrac{2}{4} =$ 2. $\dfrac{2}{3} + \dfrac{2}{3} =$

3. $\dfrac{2}{5} + \dfrac{4}{5} =$ 4. $\dfrac{4}{9} + \dfrac{7}{9} =$

5. $\dfrac{7}{10} + \dfrac{5}{10} =$ 6. $\dfrac{3}{8} + \dfrac{3}{8} + \dfrac{6}{8} =$

7. $\dfrac{1}{12} + \dfrac{7}{12} + \dfrac{6}{12} =$ 8. $\dfrac{5}{6} + \dfrac{5}{6} + \dfrac{5}{6} =$

9. $\dfrac{2}{7} + \dfrac{3}{7} + \dfrac{4}{7} =$ 10. $\dfrac{2}{3} + \dfrac{2}{3} + \dfrac{2}{3} =$

41

7.5 Mixed Numbers Can Also Be Added

If you can add whole numbers and you can add fractions, then you can add mixed numbers.

Example

The B & W Red Stamps Trading Company requires $6\frac{1}{3}$ books of stamps for a table and $2\frac{1}{3}$ books of stamps for a folding chair. How many books of stamps do you need to get both the table and the folding chair?

Solution

You must add $6\frac{1}{3} + 2\frac{1}{3}$ to find the answer to your problem. Add the fractions first and then the whole numbers. Look at the following example.

Problem	Step 1	Step 2
$6\frac{1}{3}$ $+2\frac{1}{3}$	$6\frac{1}{3}$ $+2\frac{1}{3}$ $\overline{\quad \frac{2}{3}}$	$6\frac{1}{3}$ $+2\frac{1}{3}$ $\overline{\quad 8\frac{2}{3}}$
	Add the fractions: $\frac{1}{3} + \frac{1}{3} = \frac{2}{3}$	Add the whole numbers: $6 + 2 = 8$

Thus: $6\frac{1}{3} + 2\frac{1}{3} = 8\frac{2}{3}$.

Practice Exercise 77

Add these examples.

1. $3\frac{2}{5}$
    $+1\frac{1}{5}$
    $\overline{\quad\quad}$

2. $7\frac{1}{3}$
    $+4\frac{1}{3}$
    $\overline{\quad\quad}$

3. $4\frac{2}{4}$
    $+2\frac{1}{4}$
    $\overline{\quad\quad}$

4. $\dfrac{4}{7}$

$+7\dfrac{1}{7}$

5. $4\dfrac{2}{9}$

$3\dfrac{4}{9}$

$+1\dfrac{1}{9}$

6. $13\dfrac{6}{11}$

$+\quad\dfrac{3}{11}$

7. $4\dfrac{2}{3}$

$+3\quad$ $\left(\text{Hint: } \dfrac{2}{3} + 0 = \dfrac{2}{3}\right)$

8. 5

$+2\dfrac{1}{2}$

7.6   Mixed Numbers Whose Sum May Be Simplified

If the sum of the mixed numbers is also a mixed number, sometimes the fractional part can be written in simplified form. Let's see what happens when this occurs.

Problem	Step 1	Step 2	Step 3
$4\dfrac{2}{9}$ $+2\dfrac{1}{9}$	$4\dfrac{2}{9}$ $+2\dfrac{1}{9}$ $\dfrac{3}{9}$	$4\dfrac{2}{9}$ $+2\dfrac{1}{9}$ $6\dfrac{3}{9}$	$6\dfrac{3}{9} = 6\dfrac{1}{3}$
	Add the fractions: $\dfrac{2}{9} + \dfrac{1}{9} = \dfrac{3}{9}$	Add the whole numbers: $4 + 2 = 6$	Simplify the fraction (if possible): $\dfrac{3}{9} = \dfrac{1}{3}$

Thus: $4\dfrac{2}{9} + 2\dfrac{1}{9} = 6\dfrac{1}{3}$.

Practice Exercise 78

Add the following mixed numbers, reducing all fractions to the simplest terms.

1. $5\frac{3}{8}$

    $+1\frac{1}{8}$

    _____

2. $4\frac{1}{6}$

    $+3\frac{3}{6}$

    _____

3. $23\frac{1}{12}$

    $+\ \ \frac{3}{12}$

    _____

4. $13\frac{3}{10}$

    $+\ 9\frac{2}{10}$

    _____

5. $10\frac{1}{10}$

    $9\frac{2}{10}$

    $+\ 7\frac{5}{10}$

    _____

6. $9\frac{7}{12}$

    $\frac{1}{12}$

    $+3\frac{1}{12}$

    _____

7. $7\frac{2}{8}$

    $4\frac{3}{8}$

    $+\ \ \frac{1}{8}$

    _____

8. $\frac{2}{9}$

    $4\frac{2}{9}$

    $+3\frac{2}{9}$

    _____

7.7 Mixed Numbers Whose Fractional Parts Equal One e

If the sum of the fractions is one, you add that to the whole number sum. Look at this problem:

Problem	Step 1	Step 2	Step 3
$4\frac{2}{5}$ $+3\frac{3}{5}$	$4\frac{2}{5}$ $+3\frac{3}{5}$ _____ $\frac{5}{5}$	$4\frac{2}{5}$ $+3\frac{3}{5}$ _____ $7\frac{5}{5}$	$\frac{5}{5} = 1$ $7 + 1 = 8$
	Add the fractions: $\frac{2}{5} + \frac{3}{5} = \frac{5}{5}$	Add the whole numbers: $4 + 3 = 7$	Change $\frac{5}{5}$ to 1. Add 1 to the whole number sum: $7 + 1 = 8$

Thus: $4\frac{2}{5} + 3\frac{3}{5} = 8$.

Practice Exercise 79

Add these examples.

1. $5\frac{1}{2}$

    $+2\frac{1}{2}$
    _____

2. $4\frac{3}{8}$

    $+2\frac{5}{8}$
    _____

3. $10\frac{2}{3}$

    $+\ 8\frac{1}{3}$
    _____

4. $22\frac{5}{6}$

    $+19\frac{1}{6}$
    _____

5. $3\frac{1}{7}$

    $4\frac{4}{7}$

    $+2\frac{2}{7}$
    _____

6. $\frac{4}{8}$

    $1\frac{3}{8}$

    $+6\frac{1}{8}$
    _____

7. $4\frac{4}{12}$

    $4\frac{4}{12}$

    $+4\frac{4}{12}$
    _____

8. $6\frac{3}{6}$

    $5\frac{1}{6}$

    $+\ \ \frac{2}{6}$
    _____

7.8 Mixed Numbers With Fractional Parts Whose Sum Is Greater Than One

Example

Mr. Lopez is an ironworker. He wants to cut three pieces of iron of the following lengths: $3\frac{1}{4}$ ft., $2\frac{1}{4}$ ft., and $\frac{3}{4}$ ft. How long a piece must he start with if waste from cutting is not considered?

Solution

The problem is to find the sum of $3\frac{1}{4} + 2\frac{1}{4} + \frac{3}{4}$. Here is how to solve the example.

Problem	Step 1	Step 2	Step 3
$3\frac{1}{4}$ $2\frac{1}{4}$ $+\ \frac{3}{4}$	$3\frac{1}{4}$ $2\frac{1}{4}$ $+\ \frac{3}{4}$ $\overline{\quad \frac{5}{4}}$	$3\frac{1}{4}$ $2\frac{1}{4}$ $+\ \frac{3}{4}$ $\overline{5\frac{5}{4}}$	$\frac{5}{4} = 1\frac{1}{4}$, so $5 + 1\frac{1}{4} = 6\frac{1}{4}$ ft.
	Add the fractions: $\frac{1}{4} + \frac{1}{4} + \frac{3}{4} = \frac{5}{4}$	Add the whole numbers: $3 + 2 = 5$	Since $\frac{5}{4} = 1\frac{1}{4}$, add that to the whole number 5: $5 + 1\frac{1}{4} = 6\frac{1}{4}$

Thus: $3\frac{1}{4}$ ft. $+ 2\frac{1}{4}$ ft. $+ \frac{3}{4}$ ft. $= 6\frac{1}{4}$ ft.

Practice Exercise 80

Add these mixed number problems, reducing fractions to the lowest terms.

1. $5\frac{2}{3}$
   $+2\frac{2}{3}$
   —

2. $6\frac{2}{5}$
   $+4\frac{4}{5}$
   —

3. $4\frac{4}{9}$
   $+7\frac{4}{9}$
   —

4. $8\frac{7}{10}$
   $+4\frac{5}{10}$
   —

5. $4\frac{3}{8}$
   $\frac{3}{8}$
   $+7\frac{6}{8}$
   —

6. $6\frac{1}{12}$
   $7\frac{7}{12}$
   $+1\frac{6}{12}$
   —

7. $5\frac{2}{7}$
   $\frac{3}{7}$
   $+2\frac{4}{7}$
   —

8. $\frac{2}{3}$
   $4\frac{2}{3}$
   $+5\frac{2}{3}$
   —

46

7.9 Word Problems

Before you begin, go back to the general instructions in Chapter 2, Section 2.7, to review the methods used to solve any word problem. The rules for solving a problem in addition also apply to problems in adding fractions.

Practice Exercise 81

Solve these word problems. Reduce fractions to lowest terms.

1. A recipe calls for $\frac{1}{4}$ cup of sugar now and $\frac{3}{4}$ cup of sugar later. What is the total amount of sugar used?

2. If Juan paints $\frac{2}{7}$ of the room today and $\frac{3}{7}$ of it tomorrow, how much of the room will he have painted?

3. You can get a lamp for $4\frac{1}{8}$ books of trading stamps and a table for $6\frac{5}{8}$ books of stamps. How many books of trading stamps do you need if you want both the table and the lamp?

4. Alice is a part-time hospital employee who works $6\frac{1}{4}$ hr. on Monday and $3\frac{1}{4}$ hr. on Wednesday. What is the total number of hours she works on these two days?

5. Three lengths of $4\frac{5}{6}$ yd., $10\frac{1}{6}$ yd., and 7 yd. are cut from a large ball of twine. How many yards have been removed?

6. Your employer witholds $\frac{3}{20}$ of your salary for Federal income taxes, $\frac{1}{20}$ for state taxes, and $\frac{1}{20}$ for city taxes. How much of your pay is taken out for taxes?

7. In September $2\frac{1}{16}$ in. of rain fell and in October $4\frac{5}{16}$ in. fell. How much rain fell in those two months?

8. Jack was on a diet and lost $6\frac{7}{8}$ lb. during a three-month period; he lost another $1\frac{3}{8}$ lb. during the summer. How much less does he weigh now since he began his diet?

9. Manuel needs three pieces of wood, one $16\frac{3}{16}$ in. long, a second one $5\frac{7}{16}$ in. long, and a third one $11\frac{7}{16}$ in. long. What is the total length of wood he needs?

10. In a March of Dimes Walkathon, Daisy walked $3\frac{1}{4}$ mi., Antoinette walked $5\frac{3}{4}$ mi., and Luz walked $1\frac{3}{4}$ mi. What is the total number of miles walked by the three girls?

7.10   What About Those Denominators?

All of the fractions you have added so far have had the *same* denominator. When they have the same denominator, you keep the same denominator and add only the numerators. See the following example:

$$\frac{1}{5} + \frac{2}{5} = \frac{1+2}{5} = \frac{3}{5}$$

How do you add two fractions which have *unlike* denominators?

$$\frac{2}{3} + \frac{1}{6} = ?$$

You know how to add fractions that have like denominators so let's see how you can relate this problem to the method you already know.

Can you rewrite a fraction so that it has a different denominator? These are *equivalent fractions*, which you learned about in the last chapter. To change two or more fractions into equivalent fractions you must first find the *lowest common denominator*. Review section 6.6 in Chapter 6 if it is necessary.

Remember: A common denominator is a number into which all the other denominators will divide without leaving a remainder.

Common denominators for the problem above are 6, 12, 18, 24, etc., since 3 and 6 can divide evenly into each of these numbers. You choose 6 since it is the *lowest common denominator* (L.C.D.).

Change $\frac{2}{3}$ to *sixths* by *multiplying* both the numerator and denominator by 2:

$$\frac{2}{3} = \frac{2 \times 2}{3 \times 2} = \frac{4}{6}$$

Now you add:

$$\frac{4}{6} + \frac{1}{6} = \frac{5}{6}$$

Look at the example.

Example

Mrs. Queen purchased $\frac{5}{8}$ yd. of material to make an apron. She found she needed $\frac{1}{4}$ yd. more material to make pockets and a bib. How many yards of material in all did Mrs. Queen purchase?

Solution

Problem	Step 1	Step 2	Step 3
$\frac{5}{8} + \frac{1}{4} =$	L.C.D. = 8	$\frac{5}{8} = \frac{5 \times 1}{8 \times 1} = \frac{5}{8}$ $\frac{1}{4} = \frac{1 \times 2}{4 \times 2} = \frac{2}{8}$	$\frac{5}{8} + \frac{2}{8} = \frac{7}{8}$ yd.
	The L.C.D. for the fractions is 8, since 8 is divisible by both 4 and 8.	Change each fraction into an equivalent fraction whose denominator is 8: $\frac{5}{8} = \frac{5}{8}$ $\frac{1}{4} = \frac{2}{8}$	Add the fractions: $\frac{5}{8} + \frac{2}{8} = \frac{7}{8}$

Thus: $\frac{5}{8}$ yd. $+ \frac{1}{4}$ yd. $= \frac{7}{8}$ yd.

Example

Find the sum of $\frac{5}{6}$ and $\frac{2}{5}$.

Solution

Problem	Step 1	Step 2	Step 3
$\frac{5}{6} + \frac{2}{5} =$	L.C.D. = 30	$\frac{5}{6} = \frac{5 \times 5}{6 \times 5} = \frac{25}{30}$ $\frac{2}{5} = \frac{2 \times 6}{5 \times 6} = \frac{12}{30}$	$\frac{25}{30} + \frac{12}{30} = \frac{37}{30}$ $= 1\frac{7}{30}$
	The L.C.D. for the fractions is 30. Both 6 and 5 divide evenly into 30.	Change each fraction into an equivalent fraction whose denominator is 30: $\frac{5}{6} = \frac{25}{30}$ $\frac{2}{5} = \frac{12}{30}$	Add the fractions: $\frac{25}{30} + \frac{12}{30} = \frac{37}{30}$ Simplify the improper fraction: $\frac{37}{30} = 1\frac{7}{30}$

Thus: $\frac{5}{6} + \frac{2}{5} = \frac{37}{30} = 1\frac{7}{30}$.

Practice Exercise 82

Add the following, and simplify, if possible.

1. $\frac{1}{8} + \frac{3}{4} =$

2. $\frac{1}{3} + \frac{1}{6} =$

3. $\frac{1}{7} + \frac{2}{3} =$

4. $\frac{2}{5} + \frac{5}{8} =$

5. $\frac{2}{3} + \frac{3}{5} =$

6. $\frac{3}{8} + \frac{3}{4} =$

7. $\frac{3}{4} + \frac{1}{3} =$

8. $\frac{1}{6} + \frac{3}{4} =$

9. $\frac{1}{4} + \frac{9}{10} =$

10. $\frac{3}{10} + \frac{1}{4} + \frac{2}{3} =$

11. $\frac{1}{2} + \frac{7}{8} + \frac{3}{4} =$

12. $\frac{1}{2} + \frac{2}{3} + \frac{3}{5} =$

7.11 Mixing Them All Up

Mixed numbers will not give you any trouble now that you know how to add fractions with unlike denominators. After adding the fractions, you add the whole numbers and combine the two. Look at the examples below.

Example

Mr. Carrasco finds that he needs two pieces of $\frac{1}{4}$ in. steel rod. Their lengths are $8\frac{5}{8}$ in. and $6\frac{1}{4}$ in. How much steel must he purchase?

Solution

Mr. Carrasco knew he had to find the sum of $8\frac{5}{8}$ and $6\frac{1}{4}$ to know the length of rod he needed.

Problem	Step 1	Step 2	Step 3
$8\frac{5}{8} + 6\frac{1}{4} =$	L.C.D. = 8	$8\frac{5}{8} = 8\frac{5}{8}$ $6\frac{1}{4} = 6\frac{1 \times 2}{4 \times 2} = 6\frac{2}{8}$	$8\frac{5}{8} + 6\frac{2}{8} = 14\frac{7}{8}$ in.
	Find the L.C.D. for the fractions $\frac{5}{8}$ and $\frac{1}{4}$: L.C.D. = 8	Change each fraction to the L.C.D. (8): $\frac{5}{8} = \frac{5}{8}$ $\frac{1 \times 2}{4 \times 2} = \frac{2}{8}$	Add the fractions: $\frac{5}{8} + \frac{2}{8} = \frac{7}{8}$ Add the whole numbers: $8 + 6 = 14$ Combine into a mixed number: $14\frac{7}{8}$ in.

Thus: $8\frac{5}{8}$ in. $+ 6\frac{1}{4}$ in. $= 14\frac{7}{8}$ in.

Example

Mrs. Clark needs $2\frac{5}{6}$ cups of milk to make a pudding and $1\frac{1}{3}$ cups of milk to make a cake. Find the total amount of milk required.

Solution

Problem	Step 1	Step 2	Step 3
$2\frac{5}{6} + 1\frac{1}{3} =$	L.C.D. = 6	$2\frac{5}{6} = 2\frac{5}{6}$ $1\frac{1}{3} = 1\frac{1 \times 2}{3 \times 2} =$ $1\frac{2}{6}$	$2\frac{5}{6} + 1\frac{2}{6} = 3\frac{7}{6}$ $= 3 + 1\frac{1}{6}$ $= 4\frac{1}{6}$
	The L.C.D. for $\frac{5}{6}$ and $\frac{1}{3}$ is 6.	Change each fraction to the L.C.D. (6): $\frac{5}{6} = \frac{5}{6}$ $\frac{1}{3} = \frac{1 \times 2}{3 \times 2} = \frac{2}{6}$	Add: $2\frac{5}{6} + 1\frac{2}{6} = 3\frac{7}{6}$ Change the fraction $\frac{7}{6}$ to a mixed number: $\frac{7}{6} = 1\frac{1}{6}$ Add: $3 + 1\frac{1}{6} = 4\frac{1}{6}$

Thus: $2\frac{5}{6} + 1\frac{1}{3} = 4\frac{1}{6}$.

Practice Exercise 83

Add.

1. $3\frac{1}{3} + 2\frac{1}{2} =$

2. $7\frac{1}{2} + 3\frac{3}{4} =$

3. $6\frac{3}{4} + 4\frac{2}{5} =$

4. $3\frac{3}{4} + 14\frac{7}{8} =$

5. $5\frac{3}{4} + 3\frac{1}{7} =$

6. $7\frac{5}{8} + 5\frac{9}{16} =$

7. $5\frac{3}{16} + \frac{5}{8} =$

8. $2\frac{1}{2} + 1\frac{3}{5} =$

9. $8\frac{7}{8} + 3\frac{5}{12} =$

10. $2\frac{1}{2} + \frac{2}{3} + 3\frac{3}{5} =$

11. $3\frac{2}{3} + 1\frac{1}{12} + 10\frac{1}{4} =$

7.12 More Difficult Word Problems

Before beginning the practice exercise, review the simpler problems in section 7.9. Be sure you really understand them before proceeding to these more difficult problems.

Practice Exercise 84

1. An automobile mechanic worked $3\frac{1}{3}$ hr. on Monday and $4\frac{5}{6}$ hr. on Tuesday repairing a car. How many hours had he worked on both days?

2. The distance from city A to city B is $13\frac{5}{16}$ miles. City C is $15\frac{1}{2}$ miles from city B. How far is it from city A to city C if you must pass through city B?

3. Daisy and Carmen went shopping together. Daisy bought $1\frac{1}{2}$ yd. of material for a skirt and Carmen bought $2\frac{5}{8}$ yd. of material for a dress. How many yards of material did they buy?

4. A bookcase has three shelves whose heights are $8\frac{7}{8}$ in., $6\frac{1}{4}$ in., and $7\frac{1}{2}$ in. What is the total height of the shelves?

5. In a hosiery store $3\frac{2}{3}$, $7\frac{3}{4}$, 3, and $6\frac{1}{2}$ dozen pair of stockings were sold in one day. How many dozen pairs were sold that day?

6. To complete a job, a plumber finds that he needs pipe of various lengths: $8\frac{5}{8}$ in., $10\frac{1}{2}$ in., $7\frac{3}{16}$ in., and 9 in. What is the total length of pipe needed?

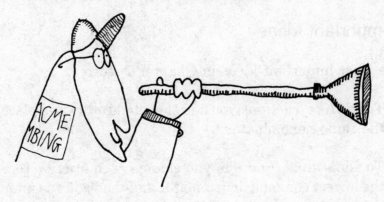

7. Find the total length of the bolt drawn below.

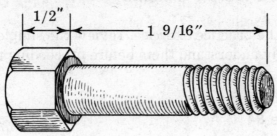

8. Find the length of the template drawn below.

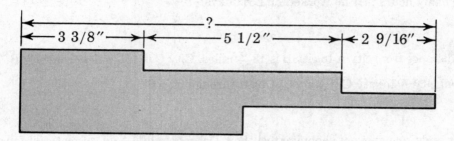

9. At the supermarket Rosa bought a $6\frac{1}{2}$ lb. pork tenderloin, an $8\frac{5}{8}$ lb. turkey, and 2 lb. of chopped meat. What was the total weight of her purchases?

10. A plot of ground is in the shape of a triangle. Find the amount of fencing needed to enclose it if the lengths of its sides are $73\frac{5}{6}$ ft., $63\frac{1}{2}$ ft., and $69\frac{1}{12}$ ft.

Term You Should Remember

Unlike fractions   Fractions whose denominators are not the same number.

Review Of Important Ideas

Some of the most important ideas in Chapter 7 were:

 To add *like fractions* you add the numerators and place the sum over the same denominator.

 To add *unlike fractions* you change each fraction to a fraction with the lowest common denominator and then add the numerators. Place the sum over the lowest common denominator.

To add *mixed numbers* you add the fractions and the whole numbers separately. Combine the sums into a mixed number.

Check What You Have Learned

This chapter covered addition of proper fractions and mixed numbers. Much of this material is probably not new to you but you might have had to review the method needed to add fractions. The posttest will let you check on your own understanding. Work the exercises carefully and try to improve your score.

Posttest 7

Add the following and write your answers in simplified form below each question.

1. $\dfrac{3}{5}$
    $+\dfrac{1}{5}$
    ____

2. $\dfrac{1}{8}$
    $+\dfrac{1}{8}$
    ____

3. $\dfrac{4}{7}$
    $+\dfrac{3}{7}$
    ____

4. $\dfrac{7}{11}$
    $+\dfrac{5}{11}$
    ____

5. $\dfrac{3}{4}$
    $+\dfrac{3}{4}$
    ____

6. $5\dfrac{1}{6}$
    $+2$
    ____

7. 8
    $+3\dfrac{1}{2}$
    ____

8. $4\dfrac{1}{5}$
    $+3\dfrac{1}{5}$
    ____

9. $3\dfrac{1}{10}$
    $+2\dfrac{3}{10}$
    ____

10. $11\dfrac{1}{8}$
    $+ 6\dfrac{7}{8}$
    ____

11. $7\dfrac{3}{5}$
    $+3\dfrac{3}{5}$
    ____

12. $9\dfrac{7}{12}$
    $+ \dfrac{11}{12}$
    ____

13. Two girls went swimming in a pool. One spent $\dfrac{5}{16}$ of an hour in the pool while the other was in the pool $\dfrac{3}{16}$ of an hour. How long were the girls in the pool?

14. Antonio was studying for his final exams. He read $14\frac{3}{4}$ pages of English, $9\frac{1}{4}$ pages of science, and $12\frac{3}{4}$ pages of math. How many pages did he study?

15. Find the least common denominator for these three fractions: $\frac{1}{4}$, $\frac{5}{8}$, $\frac{2}{3}$

16. $\begin{array}{r} \frac{1}{8} \\ +\frac{5}{16} \\ \hline \end{array}$

17. $\begin{array}{r} \frac{2}{3} \\ +\frac{7}{12} \\ \hline \end{array}$

18. $\begin{array}{r} \frac{3}{4} \\ +\frac{5}{8} \\ \hline \end{array}$

19. $\begin{array}{r} \frac{2}{5} \\ \frac{1}{2} \\ +\frac{1}{3} \\ \hline \end{array}$

20. $\begin{array}{r} 4\frac{5}{6} \\ +3\frac{2}{3} \\ \hline \end{array}$

21. $\begin{array}{r} 6\frac{5}{12} \\ 7\frac{1}{2} \\ +5\frac{2}{3} \\ \hline \end{array}$

22. How many yards of fencing are required to enclose the four-sided figure drawn below?

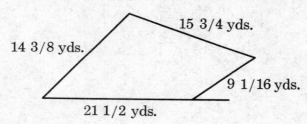

15 3/4 yds.

14 3/8 yds.

9 1/16 yds.

21 1/2 yds.

23. How many feet of wood were removed if $3\frac{3}{4}$ ft., $5\frac{1}{2}$ ft., and $1\frac{5}{12}$ ft. were cut from a 16-ft. board?

ANSWERS AND EXPLANATIONS
TO POSTTEST 7

1. $\begin{array}{r} \frac{3}{5} \\ +\frac{1}{5} \\ \hline \frac{4}{5} \end{array}$

2. $\begin{array}{r} \frac{1}{8} \\ +\frac{1}{8} \\ \hline \frac{2}{8} = \frac{1}{4} \end{array}$

3. $\begin{array}{r} \frac{4}{7} \\ +\frac{3}{7} \\ \hline \frac{7}{7} = 1 \end{array}$

56

4.

$$\frac{7}{11}$$
$$+\frac{5}{11}$$
$$\frac{12}{11} = 1\frac{1}{11}$$

5.

$$\frac{3}{4}$$
$$+\frac{3}{4}$$
$$\frac{6}{4} = 1\frac{2}{4} = 1\frac{1}{2}$$

6.

$$5\frac{1}{6}$$
$$+2$$
$$7\frac{1}{6}$$

7.

$$8$$
$$+3\frac{1}{2}$$
$$11\frac{1}{2}$$

8.

$$4\frac{1}{5}$$
$$+3\frac{1}{5}$$
$$7\frac{2}{5}$$

9.

$$3\frac{1}{10}$$
$$+2\frac{3}{10}$$
$$5\frac{4}{10} = 5\frac{2}{5}$$

10.

$$11\frac{1}{8}$$
$$+\ 6\frac{7}{8}$$
$$17\frac{8}{8} = 17 + 1$$
$$= 18$$

11.

$$7\frac{3}{5}$$
$$+3\frac{3}{5}$$
$$10\frac{6}{5} = 10 + 1\frac{1}{5}$$
$$= 11\frac{1}{5}$$

12.

$$9\frac{7}{12}$$
$$+\ \frac{11}{12}$$
$$9\frac{18}{12} = 9 + 1\frac{6}{12}$$
$$= 9 + 1\frac{1}{2}$$
$$= 10\frac{1}{2}$$

13.

$$\frac{5}{16}$$
$$+\frac{3}{16}$$
$$\frac{8}{16} = \frac{1}{2}\text{ hr.}$$

14.

$$14\frac{3}{4}$$
$$9\frac{1}{4}$$
$$+12\frac{3}{4}$$
$$35\frac{7}{4} = 35 + 1\frac{3}{4}$$
$$= 36\frac{3}{4}$$

15. L.C.D. = 24

16.

$$\frac{1}{8} = \frac{2}{16}$$
$$+\frac{5}{16} = \frac{5}{16}$$
$$\frac{7}{16}$$

17.

$$\frac{2}{3} = \frac{8}{12}$$
$$+\frac{7}{12} = \frac{7}{12}$$
$$\frac{15}{12} = 1\frac{3}{12}$$
$$= 1\frac{1}{4}$$

18.

$$\frac{3}{4} = \frac{6}{8}$$
$$+\frac{5}{8} = \frac{5}{8}$$
$$\frac{11}{8} = 1\frac{3}{8}$$

19.

$$\frac{2}{5} = \frac{12}{30}$$
$$\frac{1}{2} = \frac{15}{30}$$
$$+\frac{1}{3} = \frac{10}{30}$$
$$\frac{37}{30} = 1\frac{7}{30}$$

20.

$$4\frac{5}{6} = \frac{5}{6}$$
$$+3\frac{2}{3} = \frac{4}{6}$$
$$7\quad\frac{9}{6} = 7 + 1\frac{3}{6}$$
$$= 7 + 1\frac{1}{2}$$
$$= 8\frac{1}{2}$$

21.

$$6\frac{5}{12} = \frac{5}{12}$$
$$7\frac{1}{2} = \frac{6}{12}$$
$$+5\frac{2}{3} = \frac{8}{12}$$
$$18\quad\frac{19}{12} = 18 + 1\frac{7}{12}$$
$$= 19\frac{7}{12}$$

22. $P = 14\frac{3}{8} + 15\frac{3}{4} + 21\frac{1}{2} + 9\frac{1}{16}$

$$14\frac{3}{8} = \frac{6}{16}$$
$$15\frac{3}{4} = \frac{12}{16}$$
$$21\frac{1}{2} = \frac{8}{16}$$
$$+\ 9\frac{1}{16} = \frac{1}{16}$$
$$\overline{ 59 \qquad \frac{27}{16}} = 59 + 1\frac{11}{16}$$
$$= 60\frac{11}{16}\ \text{yds.}$$

23.
$$3\frac{3}{4} = \frac{9}{12}$$
$$5\frac{1}{2} = \frac{6}{12}$$
$$+1\frac{5}{12} = \frac{5}{12}$$
$$\overline{9 \qquad \frac{20}{12}} = 9 + 1\frac{8}{12}$$
$$= 9 + 1\frac{2}{3}$$
$$= 10\frac{2}{3}\ \text{ft.}$$

A Score of	Means That You
21–23	Did very well. You can move to Chapter 8.
18–20	Know this material except for a few points. Reread the sections about the ones you missed.
15–17	Need to check carefully on the sections you missed.
0–14	Need to review the chapter again to refresh your memory and improve your skills.

Questions	Are Covered in Section
1	7.1
2	7.2
3	7.3
4–5	7.4
6–8	7.5
9	7.6
10	7.7
11–12	7.8
13–14	7.9
15–19	7.10
20–21	7.11
22–23	7.12

ANSWERS FOR CHAPTER 7

PRETEST 7

1. $\frac{7}{8}$

2. $\frac{2}{6} = \frac{1}{3}$

3. $\frac{5}{5} = 1$

4. $\frac{11}{9} = 1\frac{2}{9}$

5. $\frac{12}{10} = 1\frac{2}{10} = 1\frac{1}{5}$

6. $7\frac{2}{3}$

7. $11\frac{1}{2}$

8. $8\frac{2}{3}$

9. $9\frac{4}{8} = 9\frac{1}{2}$

10. $19\frac{6}{6} = 20$

11. $11\frac{9}{7} = 12\frac{2}{7}$

12. $7\frac{24}{16} = 8\frac{1}{2}$

13. $\frac{3}{8} + \frac{1}{8} = \frac{4}{8} = \frac{1}{2}$

14.
$$3\frac{1}{4}$$
$$3\frac{3}{4}$$
$$+5\frac{1}{4}$$
$$\overline{11\frac{5}{4}} = 12\frac{1}{4} \text{ lb.}$$

15. L.C.D. = 12

16. $\frac{7}{8}$

17. $\frac{21}{24} = \frac{7}{8}$

18. $1\frac{1}{4}$

19. $1\frac{23}{30}$

20. $6\frac{1}{2}$

21. $16\frac{7}{16}$

22. $P = 23\frac{1}{2} + 16\frac{3}{4} + 32\frac{1}{4} + 20\frac{5}{16}$
$P = 92\frac{13}{16} \text{ ft.}$

23. $4\frac{5}{6} + 10\frac{1}{2} + 7\frac{1}{12} = 22\frac{5}{12} \text{ yds.}$

PRACTICE EXERCISE 73

1. $\frac{3}{5}$

2. $\frac{2}{3}$

3. $\frac{3}{4}$

4. $\frac{5}{7}$

5. $\frac{7}{9}$

6. $\frac{9}{11}$

7. $\frac{5}{6}$

8. $\frac{7}{8}$

9. $\frac{11}{12}$

10. $\frac{7}{10}$

PRACTICE EXERCISE 74

1. $\frac{3}{9} = \frac{1}{3}$

2. $\frac{4}{6} = \frac{2}{3}$

3. $\frac{4}{8} = \frac{1}{2}$

4. $\frac{5}{10} = \frac{1}{2}$

5. $\frac{4}{12} = \frac{1}{3}$

6. $\frac{3}{6} = \frac{1}{2}$

7. $\frac{8}{10} = \frac{4}{5}$

8. $\frac{9}{12} = \frac{3}{4}$

9. $\frac{6}{8} = \frac{3}{4}$

10. $\frac{6}{9} = \frac{2}{3}$

PRACTICE EXERCISE 75

1. $\frac{5}{5} = 1$

2. $\frac{2}{2} = 1$

3. $\frac{7}{8}$

4. $\frac{3}{3} = 1$

5. $\frac{6}{6} = 1$

6. $\frac{7}{7} = 1$

7. $\frac{8}{8} = 1$

8. $\frac{9}{10}$

9. $\frac{12}{12} = 1$

10. $\frac{6}{6} = 1$

PRACTICE EXERCISE 76

1. $\frac{5}{4} = 1\frac{1}{4}$

2. $\frac{4}{3} = 1\frac{1}{3}$

3. $\frac{6}{5} = 1\frac{1}{5}$

4. $\frac{11}{9} = 1\frac{2}{9}$

5. $\frac{12}{10} = 1\frac{2}{10} = 1\frac{1}{5}$ 6. $\frac{12}{8} = 1\frac{4}{8} = 1\frac{1}{2}$ 7. $\frac{14}{12} = 1\frac{2}{12} = 1\frac{1}{6}$ 8. $\frac{15}{6} = 2\frac{3}{6} = 2\frac{1}{2}$

9. $\frac{9}{7} = 1\frac{2}{7}$ 10. $\frac{6}{3} = 2$

PRACTICE EXERCISE 77

1. $4\frac{3}{5}$ 2. $11\frac{2}{3}$ 3. $6\frac{3}{4}$ 4. $7\frac{5}{7}$

5. $8\frac{7}{9}$ 6. $13\frac{9}{11}$ 7. $7\frac{2}{3}$ 8. $7\frac{1}{2}$

PRACTICE EXERCISE 78

1. $6\frac{4}{8} = 6\frac{1}{2}$ 2. $7\frac{4}{6} = 7\frac{2}{3}$ 3. $23\frac{4}{12} = 23\frac{1}{3}$ 4. $22\frac{5}{10} = 22\frac{1}{2}$

5. $26\frac{8}{10} = 26\frac{4}{5}$ 6. $12\frac{9}{12} = 12\frac{3}{4}$ 7. $11\frac{6}{8} = 11\frac{3}{4}$ 8. $7\frac{6}{9} = 7\frac{2}{3}$

PRACTICE EXERCISE 79

1. $7\frac{2}{2} = 8$ 2. $6\frac{8}{8} = 7$ 3. $18\frac{3}{3} = 19$ 4. $41\frac{6}{6} = 42$

5. $9\frac{7}{7} = 10$ 6. $7\frac{8}{8} = 8$ 7. $12\frac{12}{12} = 13$ 8. $11\frac{6}{6} = 12$

PRACTICE EXERCISE 80

1. $7\frac{4}{3} = 8\frac{1}{3}$ 2. $10\frac{6}{5} = 11\frac{1}{5}$ 3. $11\frac{11}{9} = 12\frac{2}{9}$ 4. $12\frac{12}{10} = 13\frac{1}{5}$

5. $11\frac{12}{8} = 12\frac{1}{2}$ 6. $14\frac{14}{12} = 15\frac{1}{6}$ 7. $7\frac{9}{7} = 8\frac{2}{7}$ 8. $9\frac{6}{3} = 11$

PRACTICE EXERCISE 81

1. $\begin{array}{r}\frac{1}{4}\\[2pt]+\frac{3}{4}\\[2pt]\hline\frac{4}{4}=1\end{array}$

2. $\begin{array}{r}\frac{2}{7}\\[2pt]+\frac{3}{7}\\[2pt]\hline\frac{5}{7}\end{array}$

3. $\begin{array}{r}4\frac{1}{8}\\[2pt]+6\frac{5}{8}\\[2pt]\hline 10\frac{6}{8}=10\frac{3}{4}\end{array}$

4. $\begin{array}{r}6\frac{1}{4}\\[2pt]+3\frac{1}{4}\\[2pt]\hline 9\frac{2}{4}=9\frac{1}{2}\ \text{hr.}\end{array}$

5. $\begin{array}{r}4\frac{5}{6}\\[2pt]10\frac{1}{6}\\[2pt]+\ 7\\[2pt]\hline 21\frac{6}{6}=22\ \text{yd.}\end{array}$

6. $\begin{array}{r}\frac{3}{20}\\[2pt]\frac{1}{20}\\[2pt]+\frac{1}{20}\\[2pt]\hline\frac{5}{20}=\frac{1}{4}\end{array}$

7.

$$2\frac{1}{16}$$
$$+4\frac{5}{16}$$
$$\overline{6\frac{6}{16}} = 6\frac{3}{8} \text{ in.}$$

8.

$$6\frac{7}{8}$$
$$+1\frac{3}{8}$$
$$\overline{7\frac{10}{8}} = 8\frac{1}{4} \text{ lb.}$$

9.

$$16\frac{3}{16}$$
$$5\frac{7}{16}$$
$$+11\frac{7}{16}$$
$$\overline{32\frac{17}{16}} = 33\frac{1}{16} \text{ in.}$$

10.

$$3\frac{1}{4}$$
$$5\frac{3}{4}$$
$$+1\frac{3}{4}$$
$$\overline{9\frac{7}{4}} = 10\frac{3}{4} \text{ mi.}$$

PRACTICE EXERCISE 82

1. $\dfrac{7}{8}$ **2.** $\dfrac{1}{2}$ **3.** $\dfrac{17}{21}$ **4.** $1\dfrac{1}{40}$

5. $1\dfrac{4}{15}$ **6.** $1\dfrac{1}{8}$ **7.** $1\dfrac{1}{12}$ **8.** $\dfrac{11}{12}$

9. $1\dfrac{3}{20}$ **10.** $1\dfrac{13}{60}$ **11.** $2\dfrac{1}{8}$ **12.** $1\dfrac{23}{30}$

PRACTICE EXERCISE 83

1. $5\dfrac{5}{6}$ **2.** $11\dfrac{1}{4}$ **3.** $11\dfrac{3}{20}$ **4.** $18\dfrac{5}{8}$

5. $8\dfrac{25}{28}$ **6.** $13\dfrac{3}{16}$ **7.** $5\dfrac{13}{16}$ **8.** $4\dfrac{1}{10}$

9. $12\dfrac{7}{24}$ **10.** $6\dfrac{23}{30}$ **11.** 15

PRACTICE EXERCISE 84

1.

$$3\frac{1}{3} = 3\frac{2}{6}$$
$$+4\frac{5}{6} = 4\frac{5}{6}$$
$$\overline{7\frac{7}{6}} = 8\frac{1}{6} \text{ hr.}$$

2.

$$13\frac{5}{16} = 13\frac{5}{16}$$
$$+15\frac{1}{2} = 15\frac{8}{16}$$
$$\overline{28\frac{13}{16}} \text{ mi.}$$

3.

$$1\frac{1}{2} = 1\frac{4}{8}$$
$$+2\frac{5}{8} = 2\frac{5}{8}$$
$$\overline{3\frac{9}{8}} = 4\frac{1}{8} \text{ yd.}$$

4.

$$8\frac{7}{8} = 8\frac{7}{8}$$
$$6\frac{1}{4} = 6\frac{2}{8}$$
$$+7\frac{1}{2} = 7\frac{4}{8}$$
$$\overline{21\frac{13}{8}} = 22\frac{5}{8} \text{ in.}$$

5.

$$3\frac{2}{3} = 3\frac{8}{12}$$
$$7\frac{3}{4} = 7\frac{9}{12}$$
$$3 = 3$$
$$+6\frac{1}{2} = 6\frac{6}{12}$$
$$\overline{19\frac{23}{12}} = 20\frac{11}{12}$$

6.

$$8\frac{5}{8} = 8\frac{10}{16}$$
$$10\frac{1}{2} = 10\frac{8}{16}$$
$$7\frac{3}{16} = 7\frac{3}{16}$$
$$+\;9 = 9$$
$$\overline{34\frac{21}{16}} = 35\frac{5}{16} \text{ in.}$$

7.
$$\frac{1}{2} = \frac{8}{16}$$
$$+1\frac{9}{16} = 1\frac{9}{16}$$
$$1\frac{17}{16} = 2\frac{1}{16} \text{ in.}$$

8.
$$3\frac{3}{8} = 3\frac{6}{16}$$
$$5\frac{1}{2} = 5\frac{8}{16}$$
$$+2\frac{9}{16} = 2\frac{9}{16}$$
$$10\frac{23}{16} = 11\frac{7}{16} \text{ in.}$$

9.
$$6\frac{1}{2} = 6\frac{4}{8}$$
$$8\frac{5}{8} = 8\frac{5}{8}$$
$$+2 = 2$$
$$16\frac{9}{8} = 17\frac{1}{8} \text{ lb.}$$

10.
$$73\frac{5}{6} = 73\frac{10}{12}$$
$$63\frac{1}{2} = 63\frac{6}{12}$$
$$+69\frac{1}{12} = 69\frac{1}{12}$$
$$205\frac{17}{12} = 206\frac{5}{12} \text{ ft.}$$

How To Study Mathematics

Before you continue to the next chapter, here are a few suggestions for studying mathematics. There are certain ideas that must be remembered, such as the meaning of symbols or terms, the formulas used, or the laws or rules that govern which process to use.

If you have not been using mathematics frequently or have not been studying recently, you need time and help to get started again. These suggestions may help you rebuild your skills in mathematics.

1. Look through any new material (a chapter or a section) to get a general idea of the topic.
2. Locate new terms or symbols and their explanations so that the use of these terms is not confusing.
3. Read explanations before you try to work any problem.
4. Learn the sequence or order of steps by working the problems as explained by the book.
5. Stop and think whether you can explain the problem. Why did you add or multiply or use whatever method you did?
6. Use practice problems as a way of checking how well you have learned the techniques. Keep trying practice problems until they become easy for you.
7. Don't let your mistakes make you angry or frustrated. Go back over what you have done to see whether your mistake was the way you set up the problem (step by step) or an error in figuring.

Subtracting fractions is the second basic operation you need to know in working with fractions. Finding the difference between two amounts is a subtraction problem. You will see that subtracting fractions is not much different from adding fractions. In fact, subtracting fractions uses much the same principles as adding fractions.

The examples which follow will test your knowledge in this type. Do them carefully.

See What You Know And Remember — Pretest 8

Do as many of these problems as you can. Some are more difficult than others. Write each answer in simplified form in the space provided.

1. $\frac{2}{3}$
   $-\frac{1}{3}$

2. $\frac{3}{4}$
   $-\frac{1}{4}$

3. $\frac{9}{8}$
   $-\frac{1}{8}$

4. $\dfrac{23}{16}$

$-\dfrac{3}{16}$

---

5. $2\dfrac{4}{5}$

$-1\dfrac{2}{5}$

---

6. $4\dfrac{5}{8}$

$-3\dfrac{3}{8}$

---

7. $8\dfrac{4}{7}$

$-\dfrac{3}{7}$

---

8. 1

$-\dfrac{2}{5}$

---

9. $5\dfrac{1}{5}$

$-\dfrac{2}{5}$

---

10. $4\dfrac{1}{6}$

$-2\dfrac{5}{6}$

---

11. $11\dfrac{1}{2}$

$-\ 2$

---

12. 9

$-4\dfrac{1}{3}$

---

13. Sonia bought $2\dfrac{1}{4}$ lb. of chopped meat. She used $1\dfrac{3}{4}$ lb. for spaghetti sauce. How many pounds of chopped meat does she have left?

14. An evening adult education class is $\dfrac{3}{4}$ of an hour long. Helen is $\dfrac{1}{4}$ of an hour late for class. How long does Helen spend in the classroom?

15. $\dfrac{2}{3}$

$-\dfrac{1}{6}$

---

16. $\dfrac{1}{2}$

$-\dfrac{1}{5}$

---

17. $3\dfrac{2}{3}$

$-2\dfrac{2}{5}$

---

18. $4\dfrac{1}{4}$

$-1\dfrac{1}{2}$

---

19. $13\dfrac{1}{8}$

$-\ 4\dfrac{3}{5}$

---

20. Maria's mother, a part-time worker, worked $31\dfrac{1}{4}$ hr. last week and $6\dfrac{2}{3}$ hr. less this week. How many hours did she work this week?

21. From the sum of $2\frac{1}{2}$ and $3\frac{3}{5}$ subtract $1\frac{1}{4}$.

22. Antonia had two pieces of fabric, one measuring $2\frac{1}{2}$ yd. and a second measuring $6\frac{3}{4}$ yd. If she used $3\frac{5}{8}$ yd. of fabric for a pants suit, how much material still remains?

23. From a roll of cloth $24\frac{7}{8}$ yd. long, a piece $12\frac{3}{4}$ yd. is cut. How many yards of material remain?

Check your answers by turning to the end of the chapter. Add the ones you had correct.

A Score of	Means That You
21–23	Did very well. You can move to Chapter 9.
18–20	Know the material except for a few points. Read the sections about the ones you missed.
15–17	Need to check carefully on the sections you missed.
0–14	Need to work with the chapter to refresh your memory and improve your skills.

Questions	Are Covered in Section
1	8.1
2	8.2
3, 4	8.3
5–7, 11	8.4
8, 12	8.5
9, 10	8.6
13, 14	8.7
15, 16	8.8
17, 21	8.9
18, 19	8.10
20, 22, 23	8.11

8.1   Subtracting Fractions

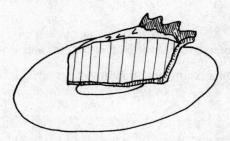

Example

Mrs. Jones baked a pie and divided it into 8 equal portions. Each portion was $\frac{1}{8}$ of the whole pie. She ate a piece for lunch, leaving $\frac{7}{8}$ of the pie. At dinner $\frac{4}{8}$ more of the pie was eaten. What part of the pie remains?

Solution

To find out the part of the pie that still remains, you must subtract the fractions:

$$\frac{7}{8} - \frac{4}{8} = ?$$

The diagram below will help you to discover how fractions are subtracted.

Part remaining   –   Part eaten at dinner   =   Part remaining
after lunch

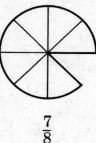   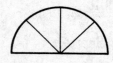

$\frac{7}{8}$ $\frac{4}{8}$ $\frac{3}{8}$

Thus: Mrs. Jones and her family ate $\frac{5}{8}$ of the pie, leaving $\frac{3}{8}$.

These fraction problems are handled exactly like the addition of fractions except that the operation is subtraction. Do you recall the names of the terms in a subtraction example?

$$\frac{7}{8} \quad \text{is the minuend}$$

$$-\frac{4}{8} \quad \text{is the subtrahend}$$

$$\frac{3}{8} \quad \text{is the difference or remainder}$$

Fractions such as $\frac{7}{8}$ and $\frac{3}{8}$ are called *like fractions* since they have the same number in the denominator, namely 8. The fractions are *eighths* or 8ths.

In subtracting like fractions, you subtract only the numerators; the denominator remains the same. Look at this example:

Problem	Step 1	Step 2
$\frac{7}{8} - \frac{4}{8} =$	$\frac{7-4}{8}$	$\frac{3}{8}$
	Subtract the numerators: $7 - 4 = 3$	Place the difference (3) over the same denominator (8): $\frac{3}{8}$

Thus: $\frac{7}{8} - \frac{4}{8} = \frac{3}{8}$.

Terms You Should Remember

Subtract   To take away.

Remainder or difference   That which is left after a part has been taken away.

Like fractions   Fractions whose denominators are the same number.

Practice Exercise 85

Find the difference for each of these problems. Follow the steps shown in the previous example.

1. $\frac{3}{5} - \frac{2}{5} =$ 2. $\frac{4}{7} - \frac{3}{7} =$ 3. $\frac{6}{11} - \frac{3}{11} =$ 4. $\frac{2}{3} - \frac{1}{3} =$

5. $\dfrac{5}{8} - \dfrac{2}{8} =$ 6. $\dfrac{3}{4} - \dfrac{2}{4} =$ 7. $\dfrac{11}{12} - \dfrac{4}{12} =$ 8. $\dfrac{9}{10} - \dfrac{2}{10} =$

9. $\dfrac{6}{6} - \dfrac{1}{6} =$ 10. $\dfrac{1}{9} - \dfrac{1}{9} =$

8.2 After Subtracting, What Next?

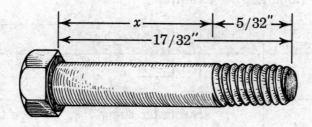

Example

Find the length x in the diagram of the bolt drawn above.

Solution

Subtracting the fractions $\dfrac{17}{32} - \dfrac{5}{32}$ will give you the answer to the example. Here is the solution:

Problem	Step 1	Step 2	Step 3
$\dfrac{17}{32} - \dfrac{5}{32} =$	$\dfrac{17 - 5}{32}$	$\dfrac{12}{32}$	$\dfrac{12 \div 4}{32 \div 4} = \dfrac{3}{8}$ in.
	Subtract the numerators: $17 - 5 = 12$	Place this difference (12) over the same denominator (32): $\dfrac{12}{32}$	Reduce the fraction by dividing each term by 4: $\dfrac{12 \div 4}{32 \div 4} = \dfrac{3}{8}$

Thus: $\dfrac{17}{32} - \dfrac{5}{32} = \dfrac{12}{32} = \dfrac{3}{8}$ in.

Practice Exercise 86

Subtract the fractions and write each difference as a reduced fraction.

1. $\dfrac{5}{6} - \dfrac{1}{6} =$

2. $\dfrac{7}{8} - \dfrac{1}{8} =$

3. $\dfrac{11}{12} - \dfrac{7}{12} =$

4. $\dfrac{11}{20} - \dfrac{3}{20} =$

5. $\dfrac{3}{4} - \dfrac{1}{4} =$

6. $\dfrac{9}{10} - \dfrac{3}{10} =$

7. $\dfrac{11}{16} - \dfrac{5}{16} =$

8. $\dfrac{19}{32} - \dfrac{15}{32} =$

9. $\dfrac{13}{14} - \dfrac{11}{14} =$

8.3 Like Fractions Whose Difference Is One Or Greater Than One

Example

Two like fractions whose difference is *one*:

$$\frac{4}{3} - \frac{1}{3} = ?$$

Solution

Problem	Step 1	Step 2	Step 3
$\dfrac{4}{3} - \dfrac{1}{3} =$	$\dfrac{4 - 1}{3}$	$\dfrac{3}{3}$	$\dfrac{3}{3} = 1$
	Subtract the numerators: $4 - 1 = 3$	Place this difference (3) over the same denominator (3): $\dfrac{3}{3}$	Any fraction whose numerator and denominator are the same number (other than 0) equals 1.

Thus: $\dfrac{4}{3} - \dfrac{1}{3} = \dfrac{3}{3} = 1$.

Example

Two like fractions whose difference is greater than *one*:

$$\frac{33}{16} - \frac{15}{16} =$$

Solution

Problem	Step 1	Step 2	Step 3
$\frac{33}{16} - \frac{15}{16} =$	$\frac{33 - 15}{16}$	$\frac{18}{16}$	$\frac{18}{16} = 1\frac{2}{16}$ $= 1\frac{1}{8}$
	Subtract the numerators: $33 - 15 = 18$	Place the difference (18) over the same denominator (16): $\frac{18}{16}$	Change the improper fraction to a mixed number: $\frac{18}{16} = 1\frac{2}{16}$ $= 1\frac{1}{8}$

Thus: $\frac{33}{16} - \frac{15}{16} = \frac{18}{16} = 1\frac{2}{16} = 1\frac{1}{8}$.

Practice Exercise 87

Subtract each of the following fractions. Write the difference in its simplest form.

1. $\frac{6}{5} - \frac{1}{5} =$

2. $\frac{5}{3} - \frac{1}{3} =$

3. $\frac{9}{4} - \frac{3}{4} =$

4. $\frac{22}{10} - \frac{11}{10} =$

5. $\frac{17}{8} - \frac{1}{8} =$

6. $\frac{13}{7} - \frac{4}{7} =$

7. $\frac{15}{12} - \frac{1}{12} =$

8. $\frac{25}{16} - \frac{5}{16} =$

9. $\frac{43}{32} - \frac{11}{32} =$

8.4 Mixed Numbers Can Also Be Subtracted

Subtracting mixed numbers is similar to adding mixed numbers. Remember, you added the whole number, then added the fractions, and combined the two sums into a *mixed* number. The same method is used for subtracting mixed numbers except that the operation is subtraction. Look at this example:

Example

Mr. Jacono, a shopkeeper, had bought $5\frac{3}{4}$ dozen shirts and sold $3\frac{1}{4}$ dozen. How many dozen still remain?

Solution

To answer the question you must subtract the two mixed numbers:

$$5\frac{3}{4} - 3\frac{1}{4}$$

Following the procedure outlined, the whole numbers and the fractions are treated individually and the result is combined. The solution follows:

Problem	Step 1	Step 2	Step 3
$5\frac{3}{4}$ $-3\frac{1}{4}$	$5\frac{3}{4}$ $-3\frac{1}{4}$ $\overline{\frac{2}{4}}$	$5\frac{3}{4}$ $-3\frac{1}{4}$ $\overline{2\frac{2}{4}}$	$2\frac{2}{4} = 2\frac{1}{2}$
	Subtract the fractions: $\frac{3-1}{4} = \frac{2}{4}$	Subtract the whole numbers: $5 - 3 = 2$	Simplify the fraction $\frac{2}{4}$: $\frac{2}{4} = \frac{1}{2}$

Thus: $5\frac{3}{4} - 3\frac{1}{4} = 2\frac{2}{4} = 2\frac{1}{2}$.

Practice Exercise 88

Subtract. Write each difference in its simplest form.

1. $6\frac{3}{7}$
$-2\frac{1}{7}$

2. $8\frac{3}{4}$
$-3\frac{1}{4}$

3. $9\frac{5}{9}$
$-\frac{1}{9}$

4. $14\frac{1}{2}$
-9

5. $4\frac{5}{6}$
$-1\frac{1}{6}$

6. $5\frac{11}{16}$
$-\frac{7}{16}$

7. $10\frac{2}{3}$
$-3\frac{1}{3}$

8. $18\frac{11}{12}$
$-7\frac{5}{12}$

9. $16\frac{5}{6}$

    $- 4$
    _____

10. $7\frac{11}{16}$

     $-3\frac{3}{16}$
     _____

8.5   Exchanging In Subtracting Fractions

Example

You use $\frac{3}{8}$ of a yard of felt to decorate your jeans. If you bought 1 yd. of felt, how much do you have left?

Solution

You must subtract the fraction $\frac{3}{8}$ from the whole number 1. The subtraction method is illustrated below.

Problem	Step 1	Step 2
1 $- \frac{3}{8}$ _____	$1 = \frac{8}{8}$ $- \frac{3}{8} = \frac{3}{8}$ _____	$1 = \frac{8}{8}$ $- \frac{3}{8} = \frac{3}{8}$ _____ $\frac{5}{8}$ yd.
	Change the whole number 1 to 8ths, since you are subtracting 8ths: $1 = \frac{8}{8}$ This is called exchanging.	Subtract the fractions: $\frac{8}{8} - \frac{3}{8} = \frac{5}{8}$

Thus: $1 \text{ yd.} - \frac{3}{8} \text{ yd.} = \frac{5}{8} \text{ yd.}$

Before you proceed to a few practice examples, let's look at another problem very similar to the preceding one.

Example

A bag of flour contains 4 cups of flour. Marilyn uses $1\frac{1}{4}$ cups of flour making a cake. How many cups of flour still remain?

72

Solution

$4 - 1\frac{1}{4} = ?$

Problem	Step 1	Step 2	Step 3
$\begin{array}{r} 4 \\ -1\frac{1}{4} \\ \hline \end{array}$	$\begin{array}{r} 4 = 3 + \frac{4}{4} \\ -1\frac{1}{4} \\ \hline \end{array}$	$\begin{array}{r} 3\frac{4}{4} \\ -1\frac{1}{4} \\ \hline \frac{3}{4} \end{array}$	$\begin{array}{r} 3\frac{4}{4} \\ -1\frac{1}{4} \\ \hline 2\frac{3}{4} \end{array}$
	Write the whole number 4 as $3 + 1$, exchanging: $1 = \frac{4}{4}$ Thus: $4 = 3 + \frac{4}{4}$	Subtract the fractions: $\frac{4-1}{4} = \frac{3}{4}$	Subtract the whole numbers: $3 - 1 = 2$

Thus: $4 - 1\frac{1}{4} = 2\frac{3}{4}$.

Practice Exercise 89

Subtract.

1. $\begin{array}{r} 1 \\ -\ \frac{7}{8} \\ \hline \end{array}$
2. $\begin{array}{r} 1 \\ -\ \frac{2}{5} \\ \hline \end{array}$
3. $\begin{array}{r} 3 \\ -\ \frac{3}{4} \\ \hline \end{array}$
4. $\begin{array}{r} 5 \\ -1\frac{1}{2} \\ \hline \end{array}$

5. $\begin{array}{r} 6 \\ -\ \frac{7}{12} \\ \hline \end{array}$
6. $\begin{array}{r} 1 \\ -\ \frac{3}{16} \\ \hline \end{array}$
7. $\begin{array}{r} 2 \\ -\ \frac{5}{8} \\ \hline \end{array}$
8. $\begin{array}{r} 13 \\ -\ 5\frac{1}{6} \\ \hline \end{array}$

9. $\begin{array}{r} 37 \\ -22\frac{3}{7} \\ \hline \end{array}$
10. $\begin{array}{r} 44 \\ -29\frac{2}{3} \\ \hline \end{array}$
11. $\begin{array}{r} 1 \\ -\ \frac{5}{32} \\ \hline \end{array}$
12. $\begin{array}{r} 37 \\ -\ 9\frac{7}{12} \\ \hline \end{array}$

8.6 Exchanging In Subtracting Mixed Numbers

Example

A truck has $4\frac{4}{5}$ gal. of antifreeze in a radiator containing $6\frac{2}{5}$ gal. of water and anti-freeze combined. How many gallons of water does the radiator hold?

Solution

$6\frac{2}{5} - 4\frac{4}{5} = ?$

Problem	Step 1	Step 2
$6\frac{2}{5}$ $-4\frac{4}{5}$	$6\frac{2}{5} = 5 + 1 + \frac{2}{5} =$ $5 + \frac{5}{5} + \frac{2}{5} = \quad 5\frac{7}{5}$ $-4\frac{4}{5} \qquad\qquad = -4\frac{4}{5}$	$5\frac{7}{5}$ $-4\frac{4}{5}$ $1\frac{3}{5}$ gal.
	Since it is impossible to subtract $\frac{4}{5}$ from $\frac{2}{5}$, rewrite the minuend $6\frac{2}{5}$ as $5 + 1 + \frac{2}{5}$. *Exchange:* $1 = \frac{5}{5}$ Thus: $5 + 1 + \frac{2}{5} =$ $5 + \frac{5}{5} + \frac{2}{5}$ Add the fractions: $5\frac{7}{5}$	Subtract the fractions: $\frac{7}{5} - \frac{4}{5} = \frac{3}{5}$ Subtract the whole numbers: $5 - 4 = 1$

Thus: $6\frac{2}{5}$ gal. $- 4\frac{4}{5}$ gal. $= 1\frac{3}{5}$ gal.

And another —

Example

$$15\frac{5}{12} - 7\frac{11}{12} = ?$$

Solution

Problem	Step 1	Step 2
$15\frac{5}{12}$ $-\ 7\frac{11}{12}$	$15\frac{5}{12} = 14 + 1 + \frac{5}{12} =$ $14 + \frac{12}{12} + \frac{5}{12} = \quad 14\frac{17}{12}$ $-\ 7\frac{11}{12} \qquad\qquad -\ 7\frac{11}{12}$	$14\frac{17}{12}$ $-\ 7\frac{11}{12}$ $7\frac{6}{12} = 7\frac{1}{2}$
	Since $\frac{11}{12}$ cannot be subtracted from $\frac{5}{12}$, rewrite the minuend $15\frac{5}{12}$ as $14 + 1 + \frac{5}{12}$. Exchange: $1 = \frac{12}{12}$ Thus: $14 + 1 + \frac{5}{12} =$ $14 + \frac{12}{12} + \frac{5}{12}$ Add the fractions: $14\frac{17}{12}$	Subtract the fractions: $\frac{17}{12} - \frac{11}{12} = \frac{6}{12}$ Subtract the whole numbers: $14 - 7 = 7$ Combine the result and simplify: $7\frac{6}{12} = 7\frac{1}{2}$

Thus: $15\frac{5}{12} - 7\frac{11}{12} = 7\frac{1}{2}$.

75

Practice Exercise 90

Subtract.

1. $4\frac{1}{3}$
   $-3\frac{2}{3}$

2. $6\frac{1}{5}$
   $-\frac{3}{5}$

3. $8\frac{3}{8}$
   $-4\frac{7}{8}$

4. $12\frac{1}{4}$
   $-3\frac{3}{4}$

5. $13\frac{2}{7}$
   $-7\frac{5}{7}$

6. $33\frac{1}{10}$
   $-9\frac{9}{10}$

7. $21\frac{3}{16}$
   $-13\frac{13}{16}$

8. $73\frac{11}{15}$
   $-68\frac{14}{15}$

9. $104\frac{17}{32}$
   $-59\frac{31}{32}$

8.7 Word Problems

Before you begin, go back to the General Instructions in Chapter 2, Section 2.7, to review the methods used to solve any word problems. The specific instructions for solving problems in subtraction (Chapter 3, Section 3.5) are the same for solving problems in subtracting fractions.

Practice Exercise 91

Solve these word problems, reducing fractions where possible.

1. Juan invests $\frac{3}{8}$ of his savings in Treasury bonds. What fraction of his savings did he not invest?

2. If a trip to Washington, D.C., takes 4 hr. by bus and only $1\frac{1}{2}$ hr. by plane, how many hours are saved in traveling by plane?

3. Lucinda weighs $114\frac{3}{4}$ lb. now, but last year she weighed $106\frac{1}{4}$ lb. How many pounds has she gained?

4. If rainfall for the month measured $2\frac{3}{8}$ in. this year and only $1\frac{7}{8}$ in. for the same month last year, then what was the increase?

5. If one machine can complete a job in $3\frac{5}{6}$ hr. while a newer machine takes only 2 hr., how many hours are saved by using the newer machine?

6. The trip to work is $12\frac{5}{9}$ mi. Mary Lou drives $7\frac{1}{9}$ mi. before running out of gas. How many miles further must she travel to work?

7. You drank $\frac{2}{8}$ of a can of soda. Mary drank $\frac{1}{8}$ of the can. What part of the soda is left?

8. You intend to study 4 hr. for a test. After studying $1\frac{5}{6}$ hr., you take a break. How many hours do you have left to study?

9. What is the length of the third side of a triangle if the sum of the other two sides is $39\frac{1}{2}$ in. and its entire perimeter is 58 in.?

10. From a 16-ft. piece of lumber a carpenter cuts three pieces measuring $4\frac{11}{16}$ ft., $5\frac{3}{16}$ ft. and $3\frac{5}{16}$ ft. How many feet still remain? (Waste from cutting should not be considered.)

8.8 Subtracting Fractions With Unlike Denominators

All of the fractions and mixed numbers you have subtracted so far have had *like denominators*. When they have the same denominator, you keep the denominator and subtract only the numerators. See the following example:

$$\frac{7}{8} - \frac{2}{8} = \frac{7-2}{8} = \frac{5}{8}$$

How do you subtract two fractions which have unlike denominators like these?

$$\frac{2}{3} - \frac{1}{6} = ?$$

You know how to subtract fractions which have like denominators, so let's see how you can relate the problem of unlike denominators to the one you already know.

First, you find the *lowest common denominator* for the two fractions. (Review Section 6.6 in Chapter 6 if it is necessary.) Second, you proceed as you did in adding fractions. Look at the method used to solve the following example.

Example

From a candle $\frac{2}{3}$ its original size, Maria burned another $\frac{1}{6}$. What part of the candle still remains?

Solution

To find the part that still remains, subtract $\frac{1}{6}$ from $\frac{2}{3}$ or

$$\frac{2}{3} - \frac{1}{6} = ?$$

Problem	Step 1	Step 2	Step 3
$\frac{2}{3}$ $-\frac{1}{6}$	L.C.D. = 6	$\frac{2 \times 2}{3 \times 2} = \frac{4}{6}$ $-\frac{1}{6} = \frac{1}{6}$	$\frac{4}{6}$ $-\frac{1}{6}$ $\frac{3}{6} = \frac{1}{2}$
	The L.C.D. for the fractions is 6, since 6 is divisible by both 3 and 6.	Change each fraction into an equivalent fraction whose denominator is 6: $\frac{2}{3} = \frac{4}{6}$ $\frac{1}{6} = \frac{1}{6}$	Subtract the fractions: $\frac{4}{6} - \frac{1}{6} =$ $\frac{4-1}{6} = \frac{3}{6}$ Simplify the fraction: $\frac{3}{6} = \frac{1}{2}$

Thus: $\frac{2}{3} - \frac{1}{6} = \frac{1}{2}$.

And another —

Example

Find the remainder when $\frac{3}{5}$ is subtracted from $\frac{2}{3}$.

Solution

$$\frac{2}{3} - \frac{3}{5} = ?$$

Problem	Step 1	Step 2	Step 3
$\begin{array}{r}\frac{2}{3}\\[4pt]-\frac{3}{5}\\ \hline\end{array}$	L.C.D. = 15	$\dfrac{2 \times 5}{3 \times 5} = \dfrac{10}{15}$ $-\dfrac{3 \times 3}{5 \times 3} = \dfrac{9}{15}$	$\dfrac{2}{3} = \dfrac{10}{15}$ $-\dfrac{3}{5} = \dfrac{9}{15}$ $\dfrac{1}{15}$
	The L.C.D. for the two fractions is 15. Both 3 and 5 divide evenly into 15.	Change each fraction into an equivalent fraction whose denominator is 15: $\dfrac{2}{3} = \dfrac{10}{15}$ $\dfrac{3}{5} = \dfrac{9}{15}$	Subtract the fractions: $\dfrac{10}{15} - \dfrac{9}{15} =$ $\dfrac{10 - 9}{15} = \dfrac{1}{15}$

Thus: $\dfrac{2}{3} - \dfrac{3}{5} = \dfrac{1}{15}$.

Practice Exercise 92

Subtract.

1. $\dfrac{1}{2} - \dfrac{2}{5} =$

2. $\dfrac{1}{2} - \dfrac{1}{4} =$

3. $\dfrac{3}{4} - \dfrac{1}{6} =$

4. $\begin{array}{r}\frac{3}{4}\\[4pt]-\frac{2}{3}\\ \hline\end{array}$

5. $\begin{array}{r}\frac{5}{8}\\[4pt]-\frac{1}{2}\\ \hline\end{array}$

6. $\begin{array}{r}\frac{3}{4}\\[4pt]-\frac{3}{5}\\ \hline\end{array}$

7. $\dfrac{5}{6} - \dfrac{2}{3} =$

8. From $\dfrac{3}{5}$ subtract $\dfrac{3}{10}$.

9. $\dfrac{13}{16} - \dfrac{1}{2} =$

10. Find the remainder when $\dfrac{3}{16}$ is subtracted from $\dfrac{7}{8}$.

8.9 Subtracting Mixed Numbers

The procedures for subtracting fractions with unlike denominators can be used in subtracting mixed numbers when the fractional parts have different denominators. After subtracting the fractional part of the mixed numbers, you then subtract the whole numbers and combine the two results. The following example will illustrate the method for you.

Example

From a piece of lumber $52\frac{1}{2}$ in. long, Mr. Gonzalez cut $24\frac{1}{8}$ in. How many inches still remain?

Solution

Subtracting $24\frac{1}{8}$ from $52\frac{1}{2}$ will result in the desired solution. Note this illustration:

Problem	Step 1	Step 2	Step 3
$52\frac{1}{2}$ $-24\frac{1}{8}$	L.C.D. = 8	$52\frac{1}{2} = 52\frac{4}{8}$ $-24\frac{1}{8} = 24\frac{1}{8}$	$52\frac{4}{8}$ $-24\frac{1}{8}$ $28\frac{3}{8}$ in.
	The L.C.D. for 2 and 8 is 8.	Change $\frac{1}{2}$ to 8ths: $\frac{1}{2} = \frac{4}{8}$	Subtract the fractions: $\frac{4}{8} - \frac{1}{8} =$ $\frac{4-1}{8} = \frac{3}{8}$ Subtract the whole numbers: $52 - 24 = 28$

Thus: $52\frac{1}{2}$ in. $- 24\frac{1}{8}$ in. $= 28\frac{3}{8}$ in.

Practice Exercise 93

Subtract and simplify.

1. $5\dfrac{5}{8}$

    $-3\dfrac{3}{16}$

2. $8\dfrac{7}{8}$

    $-3\dfrac{5}{12}$

3. $10\dfrac{7}{10}$

    $-\ 4\dfrac{1}{4}$

4. $7\dfrac{2}{3}$

    $-3\dfrac{1}{2}$

5. $6\dfrac{1}{2}$

    $-4\dfrac{3}{8}$

6. $5\dfrac{2}{5}$

    $-2\dfrac{1}{8}$

7. $4\dfrac{7}{12}$

    $-3\dfrac{1}{3}$

8. $61\dfrac{5}{8}$

    $-47\dfrac{9}{16}$

9. $156\dfrac{3}{10}$

    $-148\dfrac{1}{4}$

8.10   Exchanging In Subtracting Mixed Numbers

Example

Jack found that he now weighs $156\dfrac{1}{4}$ lb. while last year he weighed $148\dfrac{1}{2}$ lb.   How many pounds has he gained?

Solution

You must find the difference between $156\dfrac{1}{4}$ and $148\dfrac{1}{2}$ to discover the number of pounds Jack has gained:

$$156\dfrac{1}{4} - 148\dfrac{1}{2} = \ ?$$

Problem	Step 1	Step 2	Step 3
$156\frac{1}{4}$ $-148\frac{1}{2}$	$156\frac{1}{4} = 156\frac{1}{4}$ $-148\frac{1}{2} = 148\frac{2}{4}$	$156\frac{1}{4} = 155 + \frac{4}{4} + \frac{1}{4}$ $\phantom{156\frac{1}{4}} = 155\frac{5}{4}$ $-148\frac{2}{4} = 148\frac{2}{4}$	$155\frac{5}{4}$ $-148\frac{2}{4}$ $7\frac{3}{4}$ lb.
	Change each fraction to the L.C.D. (4): $\frac{1}{2} = \frac{2}{4}$	Since you can't subtract $\frac{1}{4} - \frac{2}{4} = ?$, write the minuend $156\frac{1}{4}$ as $155 + 1 + \frac{1}{4}$. Exchange: $1 = \frac{4}{4}$ Thus: $155 + 1 + \frac{1}{4} =$ $155 + \frac{4}{4} + \frac{1}{4}$ Add the fractions: $155\frac{5}{4}$	Subtract the fractions: $\frac{5}{4} - \frac{2}{4} = \frac{3}{4}$ Subtract the whole numbers: $155 - 148 = 7$

Thus: $156\frac{1}{4}$ lb. $- 148\frac{1}{2}$ lb. $= 7\frac{3}{4}$ lb.

And another —

Example

Traveling from New York to Washington, D.C., takes $5\frac{1}{2}$ hr. by car and only $3\frac{5}{6}$ hr. by train. How many fewer hours does it take traveling by train?

Solution

You must find the difference between $5\frac{1}{2}$ and $3\frac{5}{6}$ to find the answer.

Problem	Step 1	Step 2	Step 3
$5\frac{1}{2}$ $-3\frac{5}{6}$	$5\frac{1}{2} = 5\frac{3}{6}$ $-3\frac{5}{6} = 3\frac{5}{6}$	$5\frac{3}{6} = 4 + \frac{6}{6} + \frac{3}{6}$ $= 4\frac{9}{6}$ $-3\frac{5}{6} = 3\frac{5}{6}$	$4\frac{9}{6}$ $-3\frac{5}{6}$ $1\frac{4}{6} = 1\frac{2}{3}$ hr.
	Change $\frac{1}{2}$ to 6ths, the L.C.D. for both fractions: $\frac{1}{2} = \frac{3}{6}$	Since you can't subtract $\frac{3}{6} - \frac{5}{6} = ?$, write the minuend $5\frac{3}{6}$ as $4 + 1 + \frac{3}{6}$. Exchange: $1 = \frac{6}{6}$ Thus: $4 + 1 + \frac{3}{6} =$ $4 + \frac{6}{6} + \frac{3}{6}$ Add the fractions: $4\frac{9}{6}$	Subtract the fractions: $\frac{9}{6} - \frac{5}{6} = \frac{4}{6}$ Subtract the whole numbers: $4 - 3 = 1$ Simplify the fraction: $1\frac{4}{6} = 1\frac{2}{3}$

Thus: $5\frac{1}{2}$ hr. $- 3\frac{5}{6}$ hr. $= 1\frac{2}{3}$ hr.

Practice Exercise 94

Subtract and simplify.

1.　　$3\frac{1}{2}$
　　$-\frac{3}{4}$

2.　　$9\frac{7}{12}$
　　$-4\frac{3}{4}$

3.　　$8\frac{1}{5}$
　　$-2\frac{1}{2}$

4. $3\frac{1}{6}$

    $-2\frac{2}{5}$

5. $6\frac{2}{7}$

    $-5\frac{2}{3}$

6. $9\frac{1}{16}$

    $-5\frac{5}{8}$

7. $4\frac{2}{5} - 2\frac{1}{2} =$

8. $5\frac{1}{3} - \frac{1}{2} =$

9. $5\frac{3}{8} - 4\frac{2}{3} =$

8.11 More Difficult Word Problems

Look over the problems in Practice Exercise 91 before doing this exercise. Be sure you understand the previous ones thoroughly, so that you will have little difficulty with these more advanced word problems.

Practice Exercise 95

1. Mrs. Washington found that she had $\frac{7}{8}$ lb. of butter in the refrigerator. After she baked a cake which required $\frac{1}{2}$ lb. of butter, how much butter still remained?

2. From a roll of material $24\frac{1}{8}$ yd. long, a piece measuring $12\frac{3}{4}$ yd. is cut. How many yards of material remain?

3. A railroad car weighs $47\frac{7}{8}$ tons when fully loaded and $12\frac{3}{4}$ tons when empty. How many tons of cargo does it hold if it is fully loaded?

4. Anthony had completed reading $\frac{3}{5}$ of a book. By the next day $\frac{5}{8}$ of the book had been completed. What new part of the book had he read?

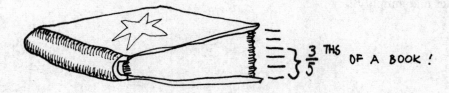

5. Beulah walked $\frac{9}{16}$ mi. while Bertha walked $\frac{2}{3}$ mi. What is the difference in the distances walked?

6. To make a single strand of jewelry chain $1\frac{5}{6}$ ft. is needed. A double strand requires $3\frac{1}{3}$ ft. of chain. How many more feet of chain are needed to make a double strand rather than a single strand?

7. From a board 16 ft. long, a carpenter cut lengths of 6 ft., $5\frac{5}{8}$ ft., $2\frac{1}{2}$ ft. and $1\frac{11}{16}$ ft. How much of the original board remains?

8. If $58\frac{1}{8}$ ft. of fencing are required to enclose a triangular plot of ground, what is the length of the third side of the triangle if the other two sides measure $19\frac{3}{16}$ ft. and $20\frac{11}{32}$ ft.?

9. Mr. Dostanza finds the dimensions of a rectangular-shaped room are $13\frac{1}{2}$ ft. by $14\frac{7}{8}$ ft. If he buys 64 ft. of molding to place around the entire room, how many feet, if any, are left over?

10. A mixture of five blends of coffee must weigh 231 lb. How many pounds of Blend E are needed if the four other blends weigh $45\frac{1}{8}$ lb., $49\frac{1}{2}$ lb., $32\frac{7}{8}$ lb., and $75\frac{3}{4}$ lb.?

Term You Should Remember

Exchanging   Replacing one thing by another.

Review Of Important Ideas

Some of the most important ideas in Chapter 8 were:

 To subtract *like fractions* you subtract the numerators and place the difference over the same denominator.

 To subtract *unlike fractions* you change each fraction to a fraction with the lowest common denominator and then subtract the numerators. Place the difference over the lowest common denominator.

 To subtract *mixed numbers* you subtract the fractions and you subtract the whole numbers separately. Combine each difference into a mixed number.

Check What You Have Learned

This chapter completed the second operation dealing with fractions, namely, the subtraction of fractions. The posttest allows you to check on your understanding of the material in the chapter. Try it to see how well you have mastered the material. Good luck!

Posttest 8

Subtract. Write each answer in its simplest form below each question.

1. $\dfrac{3}{5}$
$-\dfrac{1}{5}$
_____

2. $\dfrac{7}{10}$
$-\dfrac{1}{10}$
_____

3. $\dfrac{4}{3}$
$-\dfrac{1}{3}$
_____

4. $\dfrac{13}{8}$
$-\dfrac{3}{8}$
_____

5. $5\dfrac{2}{3}$
$-3\dfrac{1}{3}$
_____

6. $7\dfrac{3}{4}$
$-6\dfrac{1}{4}$
_____

7. $9\dfrac{5}{16}$
$-\dfrac{4}{16}$
_____

8. 1
$-\dfrac{3}{8}$
_____

9. $6\dfrac{1}{3}$
$-\dfrac{2}{3}$
_____

10. $7\dfrac{1}{8}$
$-3\dfrac{5}{8}$
_____

11. $14\dfrac{4}{7}$
-3
_____

12. 8
$-5\dfrac{3}{5}$
_____

13. Margaret bought $2\dfrac{1}{4}$ lb. of potatoes. She used $1\dfrac{3}{4}$ lb. in a stew. How many pounds of potatoes still remain?

14. A movie short subject is $\dfrac{7}{8}$ of an hour long. Donna is $\dfrac{1}{8}$ of an hour late to the theater. How much of the film (in terms of time) does she see?

15.
$$\frac{11}{12}$$
$$-\frac{3}{4}$$
———

16.
$$\frac{7}{9}$$
$$-\frac{1}{2}$$
———

17.
$$8\frac{5}{8}$$
$$-7\frac{2}{5}$$
———

18.
$$7\frac{1}{2}$$
$$-5\frac{3}{4}$$
———

19.
$$19\frac{1}{4}$$
$$-6\frac{3}{7}$$
———

20. On Saturday the Perez family traveled $236\frac{1}{4}$ mi. and on Sunday $219\frac{2}{3}$ mi. What is the difference between the two distances traveled?

21. From the sum of $4\frac{3}{8}$ and $1\frac{1}{3}$, subtract $3\frac{1}{16}$.

22. Magdalena had two packages of chopped meat, one weighing $2\frac{1}{2}$ lb. and another weighing $6\frac{3}{4}$ lb. If she used $3\frac{5}{8}$ lb. for a dish of lasagna, how many pounds still remain?

23. From a roll of wallpaper $24\frac{7}{8}$ yd. long, a piece measuring $12\frac{3}{4}$ yd. is cut. How many yards of wallpaper remain?

```
┌─────────────────────────────┐
│   ANSWERS AND EXPLANATIONS   │
│       TO POSTTEST 8          │
└─────────────────────────────┘
```

1.
$$\frac{3}{5}$$
$$-\frac{1}{5}$$
———
$$\frac{2}{5}$$

2.
$$\frac{7}{10}$$
$$-\frac{1}{10}$$
———
$$\frac{6}{10}=\frac{3}{5}$$

3.
$$\frac{4}{3}$$
$$-\frac{1}{3}$$
———
$$\frac{3}{3}=1$$

4.
$$\frac{13}{8}$$
$$-\frac{3}{8}$$
———
$$\frac{10}{8}=1\frac{2}{8}=1\frac{1}{4}$$

5.
$$5\frac{2}{3}$$
$$-3\frac{1}{3}$$
———
$$2\frac{1}{3}$$

6.
$$7\frac{3}{4}$$
$$-6\frac{1}{4}$$
———
$$1\frac{2}{4}=1\frac{1}{2}$$

7.

$$9\frac{5}{16}$$
$$-\quad\frac{4}{16}$$
$$\overline{\quad9\frac{1}{16}}$$

8.

$$1 = \frac{8}{8}$$
$$-\quad\frac{3}{8} = \frac{3}{8}$$
$$\overline{\qquad\quad\frac{5}{8}}$$

9.

$$6\frac{1}{3} = 5 + 1 + \frac{1}{3} = 5\frac{4}{3}$$
$$-\quad\frac{2}{3} \qquad\qquad\qquad = \frac{2}{3}$$
$$\overline{\qquad\qquad\qquad\qquad\quad 5\frac{2}{3}}$$

10.

$$7\frac{1}{8} = 6 + 1 + \frac{1}{8} = 6\frac{9}{8}$$
$$-3\frac{5}{8} \qquad\qquad\quad = 3\frac{5}{8}$$
$$\overline{\qquad\qquad\qquad 3\frac{4}{8} = 3\frac{1}{2}}$$

11.

$$14\frac{4}{7}$$
$$-\quad 3$$
$$\overline{\quad 11\frac{4}{7}}$$

12.

$$8\quad = 7 + 1 = 7\frac{5}{5}$$
$$-5\frac{3}{5} \qquad\quad = 5\frac{3}{5}$$
$$\overline{\qquad\qquad\quad 2\frac{2}{5}}$$

13.

$$2\frac{1}{4} = 1 + 1 + \frac{1}{4} = 1\frac{5}{4}$$
$$-1\frac{3}{4} \qquad\qquad\quad = 1\frac{3}{4}$$
$$\overline{\qquad\qquad\qquad\quad \frac{2}{4} = \frac{1}{2}\text{ lb.}}$$

14.

$$\frac{7}{8}$$
$$-\frac{1}{8}$$
$$\overline{\quad\frac{6}{8} = \frac{3}{4}\text{ hr.}}$$

15.

$$\frac{11}{12} = \frac{11}{12}$$
$$-\quad\frac{3}{4} = \frac{9}{12}$$
$$\overline{\qquad\frac{2}{12} = \frac{1}{6}}$$

16.

$$\frac{7}{9} = \frac{14}{18}$$
$$-\frac{1}{2} = \frac{9}{18}$$
$$\overline{\qquad\frac{5}{18}}$$

17.

$$8\frac{5}{8} = 8\frac{25}{40}$$
$$-7\frac{2}{5} = 7\frac{16}{40}$$
$$\overline{\qquad\quad 1\frac{9}{40}}$$

18.

$$7\frac{1}{2} = 7\frac{2}{4} = 6 + 1 + \frac{2}{4} = 6\frac{6}{4}$$
$$-5\frac{3}{4} \qquad\qquad\qquad\qquad = 5\frac{3}{4}$$
$$\overline{\qquad\qquad\qquad\qquad\qquad 1\frac{3}{4}}$$

19.

$$19\frac{1}{4} = 19\frac{7}{28} = 18 + 1 + \frac{7}{28} = 18\frac{35}{28}$$
$$-\quad 6\frac{3}{7} = \quad 6\frac{12}{28} \qquad\qquad\qquad = 6\frac{12}{28}$$
$$\overline{\qquad\qquad\qquad\qquad\qquad\qquad 12\frac{23}{28}}$$

20.

$$236\frac{1}{4} = 236\frac{3}{12} = 235 + 1 + \frac{3}{12} = 235\frac{15}{12}$$
$$-219\frac{2}{3} = 219\frac{8}{12} \qquad\qquad\qquad = 219\frac{8}{12}$$
$$\overline{\qquad\qquad\qquad\qquad\qquad\qquad 16\frac{7}{12}\text{ mi.}}$$

21.

$$4\frac{3}{8} = 4\frac{9}{24}$$
$$+1\frac{1}{3} = 1\frac{8}{24}$$
$$\overline{\qquad\quad 5\frac{17}{24}}$$

$$5\frac{17}{24} = 5\frac{34}{48}$$
$$-3\frac{1}{16} = 3\frac{3}{48}$$
$$\overline{\qquad\quad 2\frac{31}{48}}$$

22.

$$2\frac{1}{2} = 2\frac{2}{4}$$
$$+6\frac{3}{4} = 6\frac{3}{4}$$
$$\overline{\qquad\quad 8\frac{5}{4}}$$

$$8\frac{5}{4} = 8\frac{10}{8}$$
$$-3\frac{5}{3} = 3\ \frac{5}{8}$$
$$\overline{\qquad\quad 5\ \frac{5}{8}\text{ lb.}}$$

23.

$$24\frac{7}{8} = 24\frac{7}{8}$$
$$-12\frac{3}{4} = 12\frac{6}{8}$$
$$\overline{\qquad\quad 12\frac{1}{8}\text{ yd.}}$$

A Score of	Means That You
21–23	Did very well. You can move to Chapter 9.
18–20	Know this material except for a few points. Reread the sections about the ones you missed.
15–17	Need to check carefully on the sections you missed.
0–14	Need to review the chapter again to refresh your memory and improve your skills.

Questions	Are Covered in Section
1	8.1
2	8.2
3, 4	8.3
5–7, 11	8.4
8, 12	8.5
9, 10	8.6
13, 14	8.7
15, 16	8.8
17, 21	8.9
18, 19	8.10
20, 22, 23	8.11

ANSWERS FOR CHAPTER 8

PRETEST 8

1. $\frac{1}{3}$

2. $\frac{2}{4} = \frac{1}{2}$

3. $\frac{8}{8} = 1$

4. $\frac{20}{16} = 1\frac{4}{16} = 1\frac{1}{4}$

5. $1\frac{2}{5}$

6. $1\frac{2}{8} = 1\frac{1}{4}$

7. $8\frac{1}{7}$

8. $\frac{3}{5}$

9. $4\frac{4}{5}$

10. $1\frac{2}{6} = 1\frac{1}{3}$

11. $9\frac{1}{2}$

12. $4\frac{2}{3}$

13.
$$2\frac{1}{4} = 1\frac{5}{4}$$
$$-1\frac{3}{4} = 1\frac{3}{4}$$
$$\overline{\quad \frac{2}{4} = \frac{1}{2} \text{ lb.}}$$

14.
$$\frac{3}{4}$$
$$-\frac{1}{4}$$
$$\overline{\quad \frac{2}{4} = \frac{1}{2} \text{ hr.}}$$

15. $\frac{3}{6} = \frac{1}{2}$

16. $\frac{3}{10}$

17. $1\frac{4}{15}$

18. $2\frac{3}{4}$

19. $8\frac{21}{40}$

20.
$$31\frac{1}{4} = 31\frac{3}{12} = 30\frac{15}{12}$$
$$-\ 6\frac{2}{3} =\ \ 6\frac{8}{12} =\ \ 6\frac{8}{12}$$
$$\overline{\hspace{4cm}24\frac{7}{12}\ \text{hr.}}$$

21.
$$2\frac{1}{2} = 2\frac{5}{10}$$
$$+3\frac{3}{5} = 3\frac{6}{10}$$
$$\overline{\hspace{2cm}5\frac{11}{10}}$$

$$5\frac{11}{10} = 5\frac{22}{20}$$
$$-1\frac{1}{4} = 1\frac{5}{20}$$
$$\overline{\hspace{2cm}4\frac{17}{20}}$$

22.
$$2\frac{1}{2} = 2\frac{2}{4}$$
$$+6\frac{3}{4} = 6\frac{3}{4}$$
$$\overline{\hspace{2cm}8\frac{5}{4}}$$

$$8\frac{5}{4} = 8\frac{10}{8}$$
$$-3\frac{5}{8} = 3\frac{5}{8}$$
$$\overline{\hspace{2cm}5\frac{5}{8}\ \text{yd.}}$$

23.
$$24\frac{7}{8} = 24\frac{7}{8}$$
$$-12\frac{3}{4} = 12\frac{6}{8}$$
$$\overline{\hspace{2cm}12\frac{1}{8}\ \text{yd.}}$$

PRACTICE EXERCISE 85

1. $\frac{1}{5}$ 2. $\frac{1}{7}$ 3. $\frac{3}{11}$ 4. $\frac{1}{3}$ 5. $\frac{3}{8}$

6. $\frac{1}{4}$ 7. $\frac{7}{12}$ 8. $\frac{7}{10}$ 9. $\frac{5}{6}$ 10. $\frac{0}{9} = 0$

PRACTICE EXERCISE 86

1. $\frac{4}{6} = \frac{2}{3}$ 2. $\frac{6}{8} = \frac{3}{4}$ 3. $\frac{4}{12} = \frac{1}{3}$

4. $\frac{8}{20} = \frac{2}{5}$ 5. $\frac{2}{4} = \frac{1}{2}$ 6. $\frac{6}{10} = \frac{3}{5}$

7. $\frac{6}{16} = \frac{3}{8}$ 8. $\frac{4}{32} = \frac{1}{8}$ 9. $\frac{2}{14} = \frac{1}{7}$

PRACTICE EXERCISE 87

1. $\frac{5}{5} = 1$ 2. $\frac{4}{3} = 1\frac{1}{3}$ 3. $\frac{6}{4} = 1\frac{2}{4} = 1\frac{1}{2}$

4. $\frac{11}{10} = 1\frac{1}{10}$ 5. $\frac{16}{8} = 2$ 6. $\frac{9}{7} = 1\frac{2}{7}$

7. $\frac{14}{12} = 1\frac{2}{12} = 1\frac{1}{6}$ 8. $\frac{20}{16} = 1\frac{4}{16} = 1\frac{1}{4}$ 9. $\frac{32}{32} = 1$

PRACTICE EXERCISE 88

1. $4\frac{2}{7}$ 2. $5\frac{2}{4} = 5\frac{1}{2}$ 3. $9\frac{4}{9}$ 4. $5\frac{1}{2}$

5. $3\frac{4}{6} = 3\frac{2}{3}$ 6. $5\frac{4}{16} = 5\frac{1}{4}$ 7. $7\frac{1}{3}$ 8. $11\frac{6}{12} = 11\frac{1}{2}$

9. $12\frac{5}{6}$ 10. $4\frac{8}{16} = 4\frac{1}{2}$

PRACTICE EXERCISE 89

1. $\frac{1}{8}$ 2. $\frac{3}{5}$ 3. $2\frac{1}{4}$ 4. $3\frac{1}{2}$

5. $5\frac{5}{12}$ 6. $\frac{13}{16}$ 7. $1\frac{3}{8}$ 8. $7\frac{5}{6}$

9. $14\frac{4}{7}$ 10. $14\frac{1}{3}$ 11. $\frac{27}{32}$ 12. $27\frac{5}{12}$

PRACTICE EXERCISE 90

1. $\frac{2}{3}$ 2. $5\frac{3}{5}$ 3. $3\frac{4}{8} = 3\frac{1}{2}$

4. $8\frac{2}{4} = 8\frac{1}{2}$ 5. $5\frac{4}{7}$ 6. $23\frac{2}{10} = 23\frac{1}{5}$

7. $7\frac{6}{16} = 7\frac{3}{8}$ 8. $4\frac{12}{15} = 4\frac{4}{5}$ 9. $44\frac{18}{32} = 44\frac{9}{16}$

PRACTICE EXERCISE 91

1.
$$\begin{array}{r} 1 = \frac{8}{8} \\ -\frac{3}{8} = \frac{3}{8} \\ \hline \frac{5}{8} \end{array}$$

2.
$$\begin{array}{r} 4 = 3\frac{2}{2} \\ -1\frac{1}{2} = 1\frac{1}{2} \\ \hline 2\frac{1}{2}\ \text{hr.} \end{array}$$

3.
$$\begin{array}{r} 114\frac{3}{4} \\ -106\frac{1}{4} \\ \hline 8\frac{2}{4} = 8\frac{1}{2}\ \text{lb.} \end{array}$$

4.
$$\begin{array}{r} 2\frac{3}{8} = 1\frac{11}{8} \\ -1\frac{7}{8} = 1\frac{7}{8} \\ \hline \frac{4}{8} = \frac{1}{2}\ \text{in.} \end{array}$$

5.
$$\begin{array}{r} 3\frac{5}{6} \\ -2 \\ \hline 1\frac{5}{6}\ \text{hr.} \end{array}$$

6.
$$\begin{array}{r} 12\frac{5}{9} \\ -7\frac{1}{9} \\ \hline 5\frac{4}{9}\ \text{mi.} \end{array}$$

7.
$$\begin{array}{r} \frac{2}{8} \\ +\frac{1}{8} \\ \hline \frac{3}{8} \end{array} \qquad \begin{array}{r} 1 = \frac{8}{8} \\ -\frac{3}{8} = \frac{3}{8} \\ \hline \frac{5}{8} \end{array}$$

8.
$$\begin{array}{r} 4 = 3\frac{6}{6} \\ -1\frac{5}{6} = 1\frac{5}{6} \\ \hline 2\frac{1}{6}\ \text{hr.} \end{array}$$

9.
$$\begin{array}{r} 58 = 57\frac{2}{2} \\ -39\frac{1}{2} = 39\frac{1}{2} \\ \hline 18\frac{1}{2}\ \text{in.} \end{array}$$

10.
$$\begin{array}{r} 4\frac{11}{16} \\ 5\frac{3}{16} \\ +\ 3\frac{5}{16} \\ \hline 12\frac{19}{16} = 13\frac{3}{16} \end{array} \qquad \begin{array}{r} 16 = 15\frac{16}{16} \\ -13\frac{3}{16} = 13\frac{3}{16} \\ \hline 2\frac{13}{16}\ \text{ft.} \end{array}$$

PRACTICE EXERCISE 92

1. $\dfrac{1}{10}$ 2. $\dfrac{1}{4}$ 3. $\dfrac{7}{12}$ 4. $\dfrac{1}{12}$ 5. $\dfrac{1}{8}$

6. $\dfrac{3}{20}$ 7. $\dfrac{1}{6}$ 8. $\dfrac{3}{10}$ 9. $\dfrac{5}{16}$ 10. $\dfrac{11}{16}$

PRACTICE EXERCISE 93

1. $2\dfrac{7}{16}$ 2. $5\dfrac{11}{24}$ 3. $6\dfrac{9}{20}$

4. $4\dfrac{1}{6}$ 5. $2\dfrac{1}{8}$ 6. $3\dfrac{11}{40}$

7. $1\dfrac{3}{12} = 1\dfrac{1}{4}$ 8. $14\dfrac{1}{16}$ 9. $8\dfrac{1}{20}$

PRACTICE EXERCISE 94

1. $2\dfrac{3}{4}$ 2. $4\dfrac{5}{6}$ 3. $5\dfrac{7}{10}$

4. $\dfrac{23}{30}$ 5. $\dfrac{13}{21}$ 6. $3\dfrac{7}{16}$

7. $1\dfrac{9}{10}$ 8. $4\dfrac{5}{6}$ 9. $\dfrac{17}{24}$

PRACTICE EXERCISE 95

1.
$$\dfrac{7}{8} = \dfrac{7}{8}$$
$$-\dfrac{1}{2} = \dfrac{4}{8}$$
$$\dfrac{3}{8} \text{ lb.}$$

2.
$$24\dfrac{1}{8} = 24\dfrac{1}{8} = 23\dfrac{9}{8}$$
$$-12\dfrac{3}{4} = 12\dfrac{6}{8} = 12\dfrac{6}{8}$$
$$11\dfrac{3}{8} \text{ yd.}$$

3.
$$47\dfrac{7}{8} = 47\dfrac{7}{8}$$
$$-12\dfrac{3}{4} = 12\dfrac{6}{8}$$
$$35\dfrac{1}{8} \text{ tons}$$

4.
$$\dfrac{5}{8} = \dfrac{25}{40}$$
$$-\dfrac{3}{5} = \dfrac{24}{40}$$
$$\dfrac{1}{40}$$

5.
$$\dfrac{2}{3} = \dfrac{32}{48}$$
$$-\dfrac{9}{16} = \dfrac{27}{48}$$
$$\dfrac{5}{48} \text{ mi.}$$

6.
$$3\dfrac{1}{3} = 3\dfrac{2}{6} = 2\dfrac{8}{6}$$
$$-1\dfrac{5}{6} \qquad = 1\dfrac{5}{6}$$
$$1\dfrac{3}{6} = 1\dfrac{1}{2} \text{ ft.}$$

7.

$$6 = 6$$
$$5\frac{5}{8} = 5\frac{10}{16}$$
$$2\frac{1}{2} = 2\frac{8}{16}$$
$$+1\frac{11}{16} = 1\frac{11}{16}$$
$$14\frac{29}{16} = 14 + 1\frac{13}{16} = 15\frac{13}{16}$$

$$16 = 15\frac{16}{16}$$
$$-15\frac{13}{16} = 15\frac{13}{16}$$
$$\frac{3}{16}\text{ ft.}$$

8.

$$19\frac{3}{16} = 19\frac{6}{32}$$
$$+20\frac{11}{32} = 20\frac{11}{32}$$
$$39\frac{17}{32}$$

$$58\frac{1}{8} = 58\frac{4}{32} = 57\frac{36}{32}$$
$$-39\frac{17}{32} = 39\frac{17}{32}$$
$$18\frac{19}{32}\text{ ft.}$$

9.

$$13\frac{1}{2} = 13\frac{4}{8}$$
$$13\frac{1}{2} = 13\frac{4}{8}$$
$$14\frac{7}{8} = 14\frac{7}{8}$$
$$+14\frac{7}{8} = 14\frac{7}{8}$$
$$54\frac{22}{8} = 54 + 2\frac{6}{8} = 56\frac{6}{8} = 56\frac{3}{4}$$

$$64 = 63\frac{4}{4}$$
$$-56\frac{3}{4} = 56\frac{3}{4}$$
$$7\frac{1}{4}\text{ ft.}$$

10.

$$45\frac{1}{8} = 45\frac{1}{8}$$
$$49\frac{1}{2} = 49\frac{4}{8}$$
$$32\frac{7}{8} = 32\frac{7}{8}$$
$$+75\frac{3}{4} = 75\frac{6}{8}$$
$$201\frac{18}{8} = 201 + 2\frac{2}{8} = 203\frac{2}{8} = 203\frac{1}{4}$$

$$231 = 230\frac{4}{4}$$
$$-203\frac{1}{4} = 203\frac{1}{4}$$
$$27\frac{3}{4}\text{ lb.}$$

Hold It!

These problems are a general review of all that you have learned up to this point. If you get all the answers correct, you are really making progress and on your way to success in mathematics.

1.
$$386$$
$$\times\ 47$$

2. $28\overline{)483}$

3.
$$\frac{1}{2}$$
$$\frac{3}{8}$$
$$+1\frac{1}{4}$$

4. $4\frac{2}{5}$

   $-1\frac{1}{3}$

5. $98\frac{7}{8}$

   $-\frac{15}{16}$

6. $38\frac{1}{7}$

   $42\frac{20}{21}$

   $+\quad\frac{1}{3}$

7. Agnes made a two-piece suit that required $1\frac{1}{2}$ yd. for the top and $2\frac{3}{4}$ yd. for the skirt. How many yards of material did she use?

8.   4397
     2143
   $+8067$

9. John has agreed to work 12 hr. weekly. After working $8\frac{1}{8}$ hr, how many additional hours must he work?

10. When Sam joined a weight control program, he weighed $206\frac{1}{4}$ lb. During July he lost $6\frac{7}{8}$ lb. and in August he lost $5\frac{3}{4}$ lb. What was Sam's weight at the beginning of September?

ANSWERS TO "HOLD IT!"

1. 18,142

2. 17 R 7 or $17\frac{7}{28} = 17\frac{1}{4}$

3. $2\frac{1}{8}$

4. $3\frac{1}{15}$

5. $97\frac{15}{16}$

6. $81\frac{3}{7}$

7. $4\frac{1}{4}$ yd.

8. 14,607

9. $3\frac{7}{8}$ hr.

10. $193\frac{5}{8}$ lb.

Once again, remember that you should not allow your mistakes to make you angry or frustrated. Go back over the material you haven't mastered and see where your difficulty is. Keep trying and eventually you must succeed.

Mathematical Maze

This maze contains 13 mathematical words. Some are written horizontally ⟷, some vertically ↕, some diagonally ↗ ↘; some go in two or three directions. When you locate a word in the maze, draw a ring around it. The word TIMES has already been done for you. The solution is below.

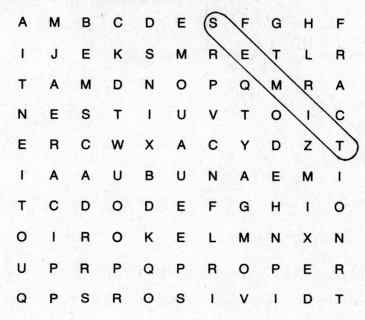

```
A   M   B   C   D   E   S   F   G   H   F
I   J   E   K   S   M   R   E   T   L   R
T   A   M   D   N   O   P   Q   M   R   A
N   E   S   T   I   U   V   T   O   I   C
E   R   C   W   X   A   C   Y   D   Z   T
I   A   A   U   B   U   N   A   E   M   I
T   C   D   O   D   E   F   G   H   I   O
O   I   R   O   K   E   L   M   N   X   N
U   P   R   P   Q   P   R   O   P   E   R
Q   P   S   R   O   S   I   V   I   D   T
```

WORDS OF:

4 letters	5 letters	6 letters	7 letters	8 letters
MODE	MIXED	PROPER	DIVISION	FRACTION
MEAN	TERMS	REDUCE	PRODUCT	QUOTIENT
AREA	TIMES	MEDIAN		

MATHEMATICAL MAZE SOLUTION

95

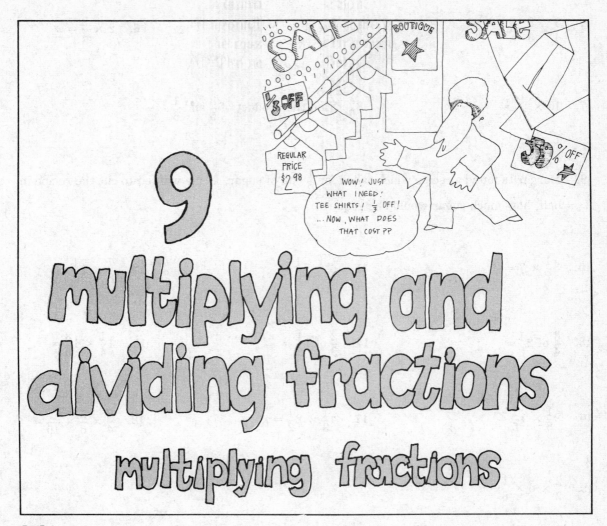

Multiplying fractions is an operation in mathematics we ordinarily use each day in our life. We see signs advertising sales of $\frac{1}{3}$ to $\frac{1}{2}$ off, $\frac{1}{2}$ price, or $\frac{3}{4}$ or $\frac{1}{2}$-acre lots for sale. Each of these statements suggests an example in multiplying fractions which will be explained in this chapter. Check your skill in multiplying fractions in the pretest which follows.

See What You Know And Remember — Pretest 9A

Do these examples as carefully as possible. Some are more difficult than others so try your best. Write the answers below each question, reducing fractions if possible.

1. $\frac{1}{5} \times \frac{2}{3} =$

2. $\frac{1}{2} \times \frac{4}{5} =$

3. $\frac{4}{7} \times \frac{5}{6} =$

4. $\frac{3}{4} \times \frac{8}{9} =$

5. $\frac{2}{3} \times 6 =$

6. $2 \times \frac{3}{4} =$

7. $10 \times \frac{9}{10} =$

8. Find the product of $\frac{3}{5}$ and $\frac{1}{6}$.

9. Margarita found a recipe which called for $\frac{1}{2}$ cup of sugar. If she wanted to cut the recipe in half, how much sugar would she require?

10. $3\frac{1}{2} \times 7 =$

11. $6 \times 7\frac{1}{2} =$

12. $\frac{7}{8} \times 1\frac{1}{4} =$

13. $\frac{3}{4}$ of $2\frac{1}{8} =$

14. $\frac{2}{9}$ of $1\frac{1}{2} =$

15. $1\frac{1}{4} \times 1\frac{3}{5} =$

16. $2\frac{1}{8} \times 1\frac{1}{5} =$

17. $4\frac{1}{3} \times 2\frac{1}{2} =$

18. $\frac{3}{4} \times \frac{5}{6} \times \frac{1}{10} =$

19. $\frac{2}{3} \times 6 \times \frac{1}{8} =$

20. Carmen rode a bicycle $3\frac{1}{2}$ mi. each day for 4 days. How many miles did she go?

21. Mary uses $2\frac{1}{2}$ cups of sugar for one cake. How much sugar will be required to bake three cakes?

22. A pattern to make an apron requires $1\frac{1}{2}$ yd. of material and a pattern to make a skirt requires $2\frac{1}{3}$ yd. of material. How much material is required if you want to make 3 aprons and 3 skirts?

Now turn to page 113 to check your answers. Add up all that you had correct.

A Score of	Means That You
20–22	Did very well. You can move to the second half of this chapter, "Dividing Fractions."
18–19	Know this material except for a few points. Read the sections about the ones you missed.
14–17	Need to check carefully on the sections you missed.
0–13	Need to work with this part of the chapter to refresh your memory and improve your skills.

Questions	Are Covered in Section
1	9.1
2–4, 8	9.2
5–7	9.3
18, 19	9.4
10–17	9.5
9, 20–22	9.6

9.1 The Product Of Two Fractions

Cookbooks have recipes which serve 4, 6, 8, or more people. A particular recipe requires $\frac{1}{3}$ teaspoon of salt for a dish serving 4 people. Suppose you wish to make this for only 2 people. Since you are cooking for half as many people, you will then cut each quantity in half. You will have to know one-half of one-third, or $\frac{1}{2}$ of $\frac{1}{3}$. This is pictured below. The rectangle is divided into three equal parts and one of these parts is $\frac{1}{3}$ of the whole.

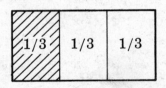

Another line is drawn horizontally, dividing the rectangle into two parts, and is shaded in the opposite direction. One of the parts is $\frac{1}{2}$ of the whole.

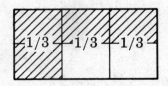

99

The rectangle is divided into 6 equal parts, one of which is crosshatched (shaded in both directions). The crosshatched section represents $\frac{1}{6}$ of the rectangle. You can now conclude that

$$\frac{1}{2} \text{ of } \frac{1}{3} = \frac{1}{6}$$

This problem can be done more quickly and efficiently by using mathematical methods:

Problem	Step 1	Step 2
$\frac{1}{2} \times \frac{1}{3} =$	$\frac{1}{2} \times \frac{1}{3} = \frac{1}{}$	$\frac{1}{2} \times \frac{1}{3} = \frac{1}{6}$
	Multiply the numerators: $1 \times 1 = 1$	Multiply the denominators: $2 \times 3 = 6$

Thus: $\frac{1}{2} \times \frac{1}{3} = \frac{1}{6}$.

Since the answer to $\frac{1}{2}$ of $\frac{1}{3}$ and $\frac{1}{2} \times \frac{1}{3}$ is the same, $\frac{1}{6}$, you can conclude that the word *of* indicates the operation of multiplication.

Practice Exercise 96

Multiply. Be sure all answers are written in their simplest form. Answers to this test and subsequent practice exercises begin on page 114.

1. $\frac{1}{3} \times \frac{1}{8} =$

2. $\frac{3}{5} \times \frac{6}{7} =$

3. $\frac{1}{2} \times \frac{1}{2} =$

4. $\frac{2}{3} \times \frac{1}{4} =$

5. $\frac{2}{5} \times \frac{5}{6} =$

6. $\frac{3}{4} \times \frac{6}{7} =$

7. $\frac{1}{2}$ of $\frac{4}{5} =$
(Write it as a multiplication example first.)

8. $\frac{1}{6} \times \frac{2}{3} =$

9. $\frac{1}{2}$ of $\frac{2}{3} =$

10. $\frac{3}{4}$ of $\frac{1}{8} =$

11. $\frac{7}{8} \times \frac{2}{3} =$

12. $\frac{2}{3} \times \frac{3}{4} =$

100

9.2 Cancellation

In the problems just completed you multiplied numerators together, multiplied denominators together, and wrote the products as a fraction. The fraction in the answer was then simplified, if possible. For example:

$$\frac{1}{3} \times \frac{3}{8} = \frac{3}{24} \text{ and } \frac{3}{24} = \frac{1}{8}$$

Another method of doing this same example would be to *cancel* (or "cross-out") any numerator or any denominator which can be divided by the same factor.

Problem	Step 1	Step 2
$\frac{1}{3} \times \frac{3}{8} =$	$\frac{1}{\cancel{3}} \times \frac{\cancel{3}}{8} =$ (with 1s)	$\frac{1}{1} \times \frac{1}{8} = \frac{1 \times 1}{1 \times 8} = \frac{1}{8}$
	Divide any numerator and any denominator by one number which will divide evenly into both. This is called *cancellation*. 3 divides into both 3s: $3 \div 3 = 1$ $3 \div 3 = 1$	Multiply the numerators: $1 \times 1 = 1$ Multiply the denominators: $1 \times 8 = 8$

Thus: $\frac{1}{3} \times \frac{3}{8} = \frac{1}{8}$.

And another —

Example

The J & R Realty Corp. was selling individual lots of $\frac{4}{9}$ acre. The Barkan family decided to purchase a lot and build a house covering $\frac{3}{4}$ of the lot. How much acreage will the house cover?

Solution

The words $\frac{3}{4}$ of the lot indicate that the problem is a multiplication example. Since the lot is $\frac{4}{9}$ of an acre, the answer to the problem will be found by the calculation,

$$\frac{3}{4} \text{ of } \frac{4}{9} \text{ or } \frac{3}{4} \times \frac{4}{9} = ?$$

This is illustrated below.

Problem	Step 1	Step 2	Step 3
$\frac{3}{4} \times \frac{4}{9} =$	$\overset{1}{\underset{1}{\frac{3}{\cancel{4}}}} \times \frac{\cancel{4}}{9} =$	$\overset{1}{\frac{\cancel{3}}{1}} \times \frac{1}{\underset{3}{\cancel{9}}} =$	$\frac{1}{1} \times \frac{1}{3} = \frac{1}{3}$
	Divide each 4 in the numerator and the denominator (*cancellation*) by 4: $4 \div 4 = 1$ $4 \div 4 = 1$	Divide the 3 in the numerator and the 9 in the denominator by 3 (*cancellation*): $3 \div 3 = 1$ $9 \div 3 = 3$	Multiply the numerators: $1 \times 1 = 1$ Multiply the denominators: $1 \times 3 = 3$

Thus: $\frac{3}{4} \times \frac{4}{9} = \frac{1}{3}$.

Practice Exercise 97

Multiply. Use the method of cancellation when you can.

1. $\frac{2}{3} \times \frac{3}{5} =$

2. $\frac{3}{4} \times \frac{2}{3} =$

3. $\frac{4}{5} \text{ of } \frac{1}{2} =$

4. $\frac{3}{8} \times \frac{10}{21} =$

5. $\frac{5}{6} \text{ of } \frac{12}{25} =$

6. $\frac{3}{5} \times \frac{7}{9} =$

7. $\frac{6}{7} \times \frac{3}{4} =$

8. $\frac{2}{5} \times \frac{5}{6} =$

9. $\frac{2}{3} \text{ of } \frac{1}{6} =$

9.3 Finding Parts Of Whole Numbers

Example

Jim went shopping for a pair of pants and saw this ad in the store window. He found a pair of pants usually selling for $8. What did they cost during the sale?

MEN'S
Big lots of Pants
1/2 PRICE

Solution

Jim must find $\frac{1}{2}$ of \$8. Recall that $\frac{1}{2}$ of \$8 is written as $\frac{1}{2} \times$ \$8. This is pictured as:

The cost of the pants during the sale is $\frac{1}{2}$ of \$8 = \$4. This problem is illustrated as:

Problem	Step 1	Step 2	Step 3
$\frac{1}{2} \times$ \$8 =	$\frac{1}{2} \times \frac{8}{1} =$	$\frac{1}{2} \times \frac{\overset{4}{\cancel{8}}}{1} = \\ 1$	$\frac{1}{1} \times \frac{4}{1} = \frac{4}{1} =$ \$4
	Write the whole number (8) as a fraction $\left(\frac{8}{1}\right)$ *Any whole number can be written as a fraction with 1 as the denominator.*	Use cancellation: $2 \div 2 = 1$ $8 \div 2 = 4$	Multiply the numerators: $1 \times 4 = 4$ Multiply the denominators: $1 \times 1 = 1$

Thus: $\frac{1}{2} \times$ \$8 = $\frac{4}{1}$ = \$4.

Whether you multiply a fraction by a whole number or a whole number by a fraction is immaterial as shown in the following example:

Example

Find the product of 10 and $\frac{2}{3}$.

Solution

Problem	Step 1	Step 2
$10 \times \frac{2}{3} =$	$\frac{10}{1} \times \frac{2}{3} =$	$\frac{10}{1} \times \frac{2}{3} = \frac{20}{3} = 6\frac{2}{3}$
	Write the whole number (10) as a fraction $\left(\frac{10}{1}\right)$.	Multiply the numerators: $10 \times 2 = 20$ Multiply the denominators: $1 \times 3 = 3$ Change $\frac{20}{3}$ into a mixed number.

Thus: $10 \times \frac{2}{3} = \frac{20}{3} = 6\frac{2}{3}$.

Practice Exercise 98

Multiply. Write each answer in its simplest form.

1. $6 \times \frac{5}{6} =$

2. Find $\frac{3}{8}$ of 50.

3. Find $\frac{2}{3}$ of 60.

4. $20 \times \frac{2}{3} =$

5. $\frac{1}{8} \times 2 =$

6. $\frac{4}{5} \times 12 =$

7. $\frac{3}{5} \times 15 =$

8. $\frac{2}{3} \times 8 =$

9. $28 \times \frac{2}{7} =$

9.4 Multiplying Three Or More Fractions

This problem isn't any more difficult than those illustrated already. Recall that to use *cancellation* you look for a number that divides evenly into any numerator and any denominator. Look at the following example to see how this method applies to multiplying three fractions.

Example

Multiply: $\frac{5}{6} \times \frac{3}{7} \times \frac{2}{5} = ?$

Solution

Problem	Step 1	Step 2	Step 3	Step 4
$\frac{5}{6} \times \frac{3}{7} \times \frac{2}{5} =$	$\frac{\overset{1}{\cancel{5}}}{6} \times \frac{3}{7} \times \frac{2}{\underset{1}{\cancel{5}}} =$	$\frac{1}{\underset{3}{\cancel{6}}} \times \frac{3}{7} \times \frac{\overset{1}{\cancel{2}}}{1} =$	$\frac{1}{\underset{1}{\cancel{3}}} \times \frac{\overset{1}{\cancel{3}}}{7} \times \frac{1}{1} =$	$\frac{1}{1} \times \frac{1}{7} \times \frac{1}{1} = \frac{1}{7}$
	Cancel: $5 \div 5 = 1$ $5 \div 5 = 1$	Cancel: $6 \div 2 = 3$ $2 \div 2 = 1$	Cancel: $3 \div 3 = 1$ $3 \div 3 = 1$	Multiply the numerators: $1 \times 1 \times 1 = 1$ Multiply the denominators: $1 \times 7 \times 1 = 7$

Thus: $\frac{5}{6} \times \frac{3}{7} \times \frac{2}{5} = \frac{1}{7}$.

Practice Exercise 99

1. $\frac{1}{2} \times \frac{2}{3} \times \frac{3}{5} =$

2. $\frac{3}{4} \times \frac{1}{2} \times \frac{8}{9} =$

3. $\frac{3}{4} \times 2 \times \frac{5}{6} =$
   (*Hint:* To solve, write the whole number 2 as a fraction $\frac{2}{1}$).

4. $\frac{7}{8} \times \frac{3}{5} \times 4 =$

9.5 Multiplying Mixed Numbers

None of the examples presented up to this point deals with mixed numbers. Multiplying mixed numbers should not cause you any difficulty if you have mastered the early portion of this chapter.

To multiply mixed numbers you change each mixed number into an improper fraction and proceed in the same way as you would to multiply any other fractions.

Example

A page measures $8\frac{7}{8}$ in. in length. How many inches are there in $\frac{1}{2}$ the page?

Solution

You must find $\frac{1}{2}$ of $8\frac{7}{8}$ in., or $\frac{1}{2} \times 8\frac{7}{8}$ in., since *of* implies multiplication. Look at the problem below.

Problem	Step 1	Step 2
$\frac{1}{2} \times 8\frac{7}{8} =$	$\frac{1}{2} \times \frac{71}{8} =$	$\frac{1 \times 71}{2 \times 8} = \frac{71}{16} = 4\frac{7}{16}$ in.
	Change the mixed number $8\frac{7}{8}$ into an improper fraction: $8\frac{7}{8} = \frac{71}{8}$	Multiply the numerators: $1 \times 71 = 71$ Multiply the denominators: $2 \times 8 = 16$ Change the improper fraction $\frac{71}{16}$ to a mixed number: $\frac{71}{16} = 4\frac{7}{16}$

Thus: $\frac{1}{2} \times 8\frac{7}{8}$ in. $= \frac{71}{16} = 4\frac{7}{16}$ in.

And another —

Example

Brass tubing costs $\$2\frac{1}{8}$ a foot. What is the cost of $1\frac{1}{3}$ ft.?

Solution

To solve this problem you must find the product:

$$\$2\frac{1}{8} \times 1\frac{1}{3} \text{ ft.} = ?$$

Problem	Step 1	Step 2	Step 3
$\$2\frac{1}{8} \times 1\frac{1}{3} =$	$\frac{17}{8} \times \frac{4}{3} =$	$\frac{17}{\overset{}{\underset{2}{8}}} \times \frac{\overset{1}{4}}{3} =$	$\frac{17}{2} \times \frac{1}{3} = \frac{17}{6} = \$2\frac{5}{6}$
	Change each mixed number into an improper fraction: $2\frac{1}{8} = \frac{17}{8}$ $1\frac{1}{3} = \frac{4}{3}$	Cancel: $8 \div 4 = 2$ $4 \div 4 = 1$	Multiply the numerators: $17 \times 1 = 17$ Multiply the denominators: $2 \times 3 = 6$ Change the improper fraction to a mixed number: $\frac{17}{6} = 2\frac{5}{6}$

Thus: $\$2\frac{1}{8} \times 1\frac{1}{3}$ ft. $= \frac{17}{6} = \$2\frac{5}{6}$.

Practice Exercise 100

Multiply. Write each answer in its simplest form.

1. $\frac{4}{5} \times 8\frac{3}{4} =$

2. $2\frac{3}{5} \times 5 =$

3. $5\frac{1}{3} \times 3\frac{1}{4} =$

4. $2\frac{1}{2} \times 4\frac{1}{4} =$

5. $2\frac{2}{5} \times \frac{2}{3} =$

6. $\frac{1}{4} \times 4\frac{1}{5} =$

7. $2\frac{2}{3} \times 12 =$

8. $3\frac{1}{2} \times 4\frac{2}{3} =$

9. $7\frac{1}{5} \times 6\frac{3}{4} =$

10. $1\frac{2}{3} \times 4\frac{1}{5} =$

11. $3\frac{1}{2} \times 5\frac{1}{2} =$

12. $3\frac{1}{4} \times 4\frac{1}{5} =$

9.6 Word Problems

Before we begin word problems, let's renew your skills in mathematics. As you work on problems you have probably noted that you need a system or way of "setting up" the problem. Here is one system that has been divised; see if it helps you.

1. Read the problem through to get the main idea.
2. Reread the problem for details:
   a. What terms or symbols are used?
   b. What facts are given?
   c. What is to be found?
3. Decide the process or method that will solve the problem.
4. Work the problem.
5. Check that your answer is logical and correct.

Use your knowledge and the above suggestions to solve this problem:

Example

At $\$\frac{3}{4}$ per sq. ft., what is the cost of papering a wall whose dimensions are $8\frac{1}{4}$ ft. $\times$ $7\frac{1}{3}$ ft.?

Solution

1. The main idea of the problem is *area* (review Chapter 4, Section 4.8).
2. a. The key symbols and terms for a wall are rectangle, length, width, and the formula for area.
   b. The facts given are that the length is $8\frac{1}{4}$ ft., the width is $7\frac{1}{3}$ ft., and the price per sq. ft. is $\$\frac{3}{4}$.
   c. The unknown facts are the number of square feet (area) the wall contains and the cost of papering the wall.
3. In this problem, you require the formula for the area of the rectangle.
   Area = Length $\times$ Width

   Area = $8\frac{1}{4}$ ft. $\times$ $7\frac{1}{3}$ ft.

4. When you work the problem, you must know how to multiply fractions or mixed numbers.

   I. Area = $8\frac{1}{4} \times 7\frac{1}{3}$

   Area = $\frac{33}{4} \times \frac{22}{3}$

$$\text{Area} = \frac{\overset{11}{\cancel{33}}}{\underset{2}{\cancel{4}}} \times \frac{\overset{11}{\cancel{22}}}{\underset{1}{\cancel{3}}}$$

$$\text{Area} = \frac{121}{2}$$

$$\text{Area} = 60\frac{1}{2} \text{ sq. ft.}$$

II. Cost = number of sq. ft. × price per sq. ft.

$$\text{Cost} = 60\frac{1}{2} \times \$\frac{3}{4}$$

$$\text{Cost} = \frac{121}{2} \times \frac{3}{4}$$

$$\text{Cost} = \frac{363}{8}$$

$$\text{Cost} = \$45\frac{3}{8}$$

5. The answer is logical because the area is approximately 60 sq. ft. and the cost is less than $1 per sq. ft. Thus our result should be less than $60, which it is.

This illustrates one method you can use to solve a problem. Another approach may be to set up a diagram that fits the example and use the diagram to help you solve the problem. There are many different methods to employ when you are faced with a problem and the need for a solution. Try your skills on the practice exercise which follows.

Practice Exercise 101

1. A case of soda contains 24 bottles. If Mrs. Smith ordered $\frac{2}{3}$ of the case, how many bottles will she receive?

2. The trip to work each day and back is $8\frac{7}{8}$ mi. If Mr. Goris does this 5 days a week, how many miles does he travel?

3. Mr. Clark had to cut 24 pieces of wood, each $7\frac{1}{16}$ in. long, from a 16 ft. board. How many inches remained after the pieces were cut?   (*Hint:* You must convert 16 feet to inches.)

4. Mr. and Mrs. Young decide to save a portion of their salaries for next year's vacation trip. They find that they can save only $\frac{1}{10}$ of their combined weekly salaries of $225.

a. How much do they save each week?

b. How much will they save at the end of a year (52 weeks = 1 year)?

5. If $3\frac{7}{8}$ yd. of material are needed to make a dress, how many yards will be needed to complete an order for 300 dresses?

6. Find the area of a rectangle whose dimensions are $6\frac{3}{5}$ ft. by $8\frac{4}{11}$ ft.

7. Sam can do a certain job in $3\frac{3}{4}$ hr. while John can do the same job in $\frac{1}{3}$ of that time. How many hours does it take John to do the job?

8. Find the area of a triangle whose base is $4\frac{5}{8}$ in. and whose height is $9\frac{1}{2}$ in. The formula for the area of a triangle is $A = \frac{1}{2}b \times h$ where A represents area, b represents the base, and h the height.

9. Mrs. Jackson made 36 cupcakes for a church dinner. Some of the cupcakes were chocolate and some vanilla. If $\frac{2}{3}$ of them were vanilla, how many chocolate cupcakes were there?

10. If $2\frac{7}{8}$ yd. of material is needed for a dress and $2\frac{1}{16}$ yd. for a blouse, then how much material is required if Jane makes 2 dresses and 4 blouses?

Terms You Should Remember

Cancellation   The removal of a common factor from the numerator and denominator of a fraction.

Area   The measure of a flat, open surface or space.

Product   The result obtained in a multiplication example.

110

Review Of Important Ideas

Some of the most important ideas thus far in Chapter 9 have been:

 To multiply fractions, you multiply the numerators, multiply the denominators, and write the result as a fraction.

 The word *of* indicates the operation of multiplication. For example: two-thirds *of* 6 is $\frac{2}{3} \times 6$.

 Cancellation is used when multiplying fractions if a numerator and a denominator contain a common factor.

 Finding the area of a geometric figure requires you to learn the area formula associated with the figure. For example: $A = L \times W$ is the area formula for a rectangle.

Check What You Have Learned

This test gives you a chance to see if you have been successful in multiplying fractions. Try hard for a satisfactory grade.

Posttest 9A

Write your answers below each question.

1. $\frac{2}{5} \times \frac{1}{3} =$

2. $\frac{1}{3} \times \frac{6}{7} =$

3. $\frac{5}{8} \times \frac{6}{7} =$

4. $\frac{3}{5} \times \frac{10}{21} =$

5. $\frac{3}{4} \times 8 =$

6. $3 \times \frac{5}{6} =$

7. $12 \times \frac{7}{12} =$

8. Find the product of $\frac{4}{5}$ and $\frac{1}{12}$.

9. Solange's favorite recipe calls for $\frac{1}{3}$ cup of sugar. If she wanted to cut the recipe in thirds, how much sugar would she require?

10. $4\frac{1}{2} \times 9 =$ 11. $10 \times 6\frac{1}{2} =$ 12. $\frac{5}{6} \times 1\frac{1}{6} =$

13. $\frac{2}{3}$ of $2\frac{1}{8} =$ 14. $\frac{2}{21} \times 3\frac{1}{2} =$ 15. $2\frac{1}{4} \times 1\frac{7}{9} =$

16. $3\frac{1}{4} \times 1\frac{1}{5} =$ 17. $3\frac{1}{3} \times 2\frac{1}{3} =$ 18. $\frac{2}{3} \times \frac{5}{6} \times \frac{1}{15} =$

19. $\frac{2}{5} \times 10 \times \frac{1}{8} =$

20. Daisy walks $1\frac{1}{2}$ mi. 6 days a week to the train station. How many miles a week does she walk?

21. Nadia wishes to bake 3 apple pies. If $3\frac{1}{2}$ lb. of apples are used for each pie, how many pounds in all are needed?

22. Mike needs $1\frac{1}{2}$ ft. of steel tubing to make a toy car and $2\frac{1}{4}$ ft. to make a toy train. How many feet of steel tubing are needed if Mike makes 4 toy cars and 5 toy trains?

ANSWERS AND EXPLANATIONS
TO POSTTEST 9A

1. $\frac{2}{5} \times \frac{1}{3} = \frac{2}{15}$ 2. $\frac{1}{\cancel{3}} \times \frac{\cancel{6}^{\,2}}{7} = \frac{2}{7}$ 3. $\frac{5}{\cancel{8}} \times \frac{\cancel{6}^{\,3}}{7} = \frac{15}{28}$

4. $\frac{\cancel{8}^{\,1}}{\cancel{5}_1} \times \frac{\cancel{10}^{\,2}}{\cancel{21}_7} = \frac{2}{7}$ 5. $\frac{3}{\cancel{4}_1} \times \frac{\cancel{8}^{\,2}}{1} = \frac{6}{1} = 6$ 6. $\frac{\cancel{2}^{\,1}}{1} \times \frac{5}{\cancel{6}_2} = \frac{5}{2} = 2\frac{1}{2}$

7. $\frac{\cancel{12}^{\,1}}{1} \times \frac{7}{\cancel{12}_1} = \frac{7}{1} = 7$ 8. $\frac{\cancel{4}^{\,1}}{5} \times \frac{1}{\cancel{12}_3} = \frac{1}{15}$ 9. $\frac{1}{3} \times \frac{1}{3} = \frac{1}{9}$ cup

10. $\frac{9}{2} \times \frac{9}{1} = \frac{81}{2} = 40\frac{1}{2}$ 11. $\frac{10}{1} \times \frac{13}{\cancel{2}_1}^{\,5} = \frac{65}{1} = 65$ 12. $\frac{5}{6} \times \frac{7}{6} = \frac{35}{36}$

13. $\dfrac{\overset{1}{\cancel{2}}}{3} \times \dfrac{17}{\underset{4}{\cancel{8}}} = 1\dfrac{5}{12}$

14. $\dfrac{\overset{1}{\cancel{2}}}{\underset{3}{\cancel{21}}} \times \dfrac{\overset{1}{\cancel{7}}}{\underset{1}{\cancel{2}}} = \dfrac{1}{3}$

15. $\dfrac{\overset{1}{\cancel{8}}}{\underset{1}{\cancel{4}}} \times \dfrac{\overset{4}{\cancel{16}}}{\underset{1}{\cancel{8}}} = 4$

16. $\dfrac{13}{\underset{2}{\cancel{4}}} \times \dfrac{\overset{3}{\cancel{6}}}{5} = 3\dfrac{9}{10}$

17. $\dfrac{10}{3} \times \dfrac{7}{3} = \dfrac{70}{9} = 7\dfrac{7}{9}$

18. $\dfrac{\overset{1}{\cancel{2}}}{3} \times \dfrac{\overset{1}{\cancel{5}}}{\underset{3}{\cancel{8}}} \times \dfrac{1}{\underset{3}{\cancel{15}}} = \dfrac{1}{27}$

19. $\dfrac{\overset{1}{\cancel{2}}}{\underset{1}{\cancel{5}}} \times \dfrac{\overset{\overset{1}{\cancel{2}}}{\cancel{10}}}{1} \times \dfrac{1}{\underset{\underset{2}{\cancel{4}}}{\cancel{8}}} = \dfrac{1}{2}$

20. $6 \times 1\dfrac{1}{2} = \dfrac{\overset{3}{\cancel{6}}}{1} \times \dfrac{3}{\underset{1}{\cancel{2}}} = 9 \text{ mi.}$

21. $3 \times 3\dfrac{1}{2} = \dfrac{3}{1} \times \dfrac{7}{2} = \dfrac{21}{2} = 10\dfrac{1}{2} \text{ lb.}$

22. $4 \times 1\dfrac{1}{2} = \dfrac{\overset{2}{\cancel{4}}}{1} \times \dfrac{3}{\underset{1}{\cancel{2}}} = 6$

$5 \times 2\dfrac{1}{4} = \dfrac{5}{1} \times \dfrac{9}{4} = \dfrac{45}{4} = 11\dfrac{1}{4}$

$$\begin{array}{r} 6 \\ + 11\dfrac{1}{4} \\ \hline 17\dfrac{1}{4} \text{ ft.} \end{array}$$

A Score of	Means That You
20–22	Did very well. You can move to the next part of this chapter, "Dividing Fractions."
18–19	Know this material except for a few points. Reread the sections about the ones you missed.
14–17	Need to check carefully on the sections you missed.
0–13	Need to review this part of the chapter again to refresh your memory and improve your skills.

Questions	Are Covered in Section
1	9.1
2–4, 8	9.2
5–7	9.3
18, 19	9.4
10–17	9.5
9, 20–22	9.6

ANSWERS FOR CHAPTER 9 — A

PRETEST 9A

1. $\dfrac{2}{15}$ 2. $\dfrac{2}{5}$ 3. $\dfrac{10}{21}$ 4. $\dfrac{2}{3}$

5. 4   6. $1\frac{1}{2}$ 7. 9   8. $\frac{1}{10}$

9. $\frac{1}{4}$ cup   10. $24\frac{1}{2}$ 11. 45   12. $1\frac{3}{32}$

13. $1\frac{19}{32}$ 14. $\frac{1}{3}$ 15. 2   16. $2\frac{11}{20}$

17. $10\frac{5}{6}$ 18. $\frac{1}{16}$ 19. $\frac{1}{2}$ 20. 14 mi.

21. $7\frac{1}{2}$ cups   22. $11\frac{1}{2}$ yd.

PRACTICE EXERCISE 96

1. $\frac{1}{24}$ 2. $\frac{18}{35}$ 3. $\frac{1}{4}$ 4. $\frac{1}{6}$

5. $\frac{1}{3}$ 6. $\frac{9}{14}$ 7. $\frac{2}{5}$ 8. $\frac{1}{9}$

9. $\frac{1}{3}$ 10. $\frac{3}{32}$ 11. $\frac{7}{12}$ 12. $\frac{1}{2}$

PRACTICE EXERCISE 97

1. $\frac{2}{5}$ 2. $\frac{1}{2}$ 3. $\frac{2}{5}$ 4. $\frac{5}{28}$ 5. $\frac{2}{5}$

6. $\frac{7}{15}$ 7. $\frac{9}{14}$ 8. $\frac{1}{3}$ 9. $\frac{1}{9}$

PRACTICE EXERCISE 98

1. 5   2. $\frac{75}{4} = 18\frac{3}{4}$ 3. 40   4. $\frac{40}{3} = 13\frac{1}{3}$ 5. $\frac{1}{4}$

6. $\frac{48}{5} = 9\frac{3}{5}$ 7. 9   8. $\frac{16}{3} = 5\frac{1}{3}$ 9. 8

PRACTICE EXERCISE 99

1. $\frac{1}{5}$ 2. $\frac{1}{3}$ 3. $\frac{5}{4} = 1\frac{1}{4}$ 4. $\frac{21}{10} = 2\frac{1}{10}$

PRACTICE EXERCISE 100

1. 7   2. 13   3. $17\frac{1}{3}$ 4. $10\frac{5}{8}$

5. $1\frac{3}{5}$ 6. $1\frac{1}{20}$ 7. 32   8. $16\frac{1}{3}$

9. $48\frac{3}{5}$ 10. 7   11. $19\frac{1}{4}$ 12. $13\frac{13}{20}$

114

1. $\frac{2}{3} \times 24 =$

$$\frac{2}{\cancel{3}} \times \frac{\overset{8}{\cancel{24}}}{1} = 16$$

2. $5 \times 8\frac{7}{8} =$

$$\frac{5}{1} \times \frac{71}{8} = \frac{355}{8} = 44\frac{3}{8} \text{ mi.}$$

3. $24 \times 7\frac{1}{16} =$

$$\frac{24}{1} \times \frac{113}{16} = \frac{339}{2} = 169\frac{1}{2} \text{ in.}$$

16 ft. = 192 in.

$$\begin{array}{r} -169\frac{1}{2} \\ \hline 22\frac{1}{2} \text{ in.} \end{array}$$

4. a. $\frac{1}{\cancel{10}} \times \frac{\overset{45}{\cancel{225}}}{1} = \frac{45}{2} = \$22\frac{1}{2}$

b. $52 \times \$22\frac{1}{2} =$

$$\frac{\overset{26}{\cancel{52}}}{1} \times \frac{45}{\cancel{2}} = \$1,170$$

5. $300 \times 3\frac{7}{8} =$

$$\frac{\overset{75}{\cancel{300}}}{1} \times \frac{31}{\cancel{8}} = \frac{2325}{2} = 1,162\frac{1}{2} \text{ yds.}$$

6. $A = L \times W$

$A = 6\frac{3}{5} \times 8\frac{4}{11}$

$$= \frac{\overset{3}{\cancel{33}}}{5} \times \frac{92}{\cancel{11}} = \frac{276}{5} = 55\frac{1}{5} \text{ sq. ft.}$$

7. $\frac{1}{3} \times 3\frac{3}{4} =$

$$\frac{1}{\cancel{3}} \times \frac{\overset{5}{\cancel{15}}}{4} = \frac{5}{4} = 1\frac{1}{4} \text{ hrs.}$$

8. $A = \frac{1}{2}b \times h$

$= \frac{1}{2} \times 4\frac{5}{8} \times 9\frac{1}{2}$

$= \frac{1}{2} \times \frac{37}{8} \times \frac{19}{2}$

$A = \frac{703}{32} = 21\frac{31}{32} \text{ sq. ft.}$

9. $\frac{2}{\cancel{3}} \times \frac{\overset{12}{\cancel{36}}}{1} = 24 \text{ vanilla}$

$$\begin{array}{r} 36 \\ -24 \\ \hline 12 \text{ chocolate} \end{array}$$

10. $2 \times 2\frac{7}{8} =$

$$\frac{\overset{1}{\cancel{2}}}{1} \times \frac{23}{\cancel{8}} = \frac{23}{4} = 5\frac{3}{4}$$

$4 \times 2\frac{1}{16} =$

$$\frac{\overset{1}{\cancel{4}}}{1} \times \frac{33}{\cancel{16}} = \frac{33}{4} = 8\frac{1}{4}$$

$$\begin{array}{r} 5\frac{3}{4} \\ +8\frac{1}{4} \\ \hline 13\frac{4}{4} = 14 \text{ yd.} \end{array}$$

dividing fractions

Division is our last operation dealing with fractions. It requires the same intensive study you have been doing for the other operations. Division of fractions will help you to estimate the number of floor tiles, each measuring $\frac{9}{16}$ sq. ft., needed to cover a floor with dimensions of 9 ft. × 12 ft. or any other size. It will allow you to discover how many trips you need to make to remove $7\frac{1}{2}$ tons of gravel if your truck holds only $2\frac{1}{2}$ tons. It is an important operation with many useful applications.

Do you still have your skill in dividing fractions? Here's your chance to find out. Take the pretest which follows.

See What You Know And Remember — Pretest 9B

Work these examples as carefully as possible. Some are more difficult than others. Write the answers, in simplified form, below each question.

1. $9 \div \frac{1}{3} =$

2. $20 \div \frac{4}{5} =$

3. $5 \div \frac{3}{8} =$

4. $18 \div 2\frac{1}{4} =$

5. $4 \div 2\frac{5}{8} =$

6. $6 \div 6\frac{3}{4} =$

7. $15 \div 1\frac{9}{16} =$

8. $\frac{3}{4} \div \frac{2}{3} =$

9. $\frac{5}{8} \div \frac{15}{16} =$

10. $\frac{7}{8} \div \frac{7}{32} =$

11. $\frac{2}{3} \div \frac{5}{12} =$

12. $\frac{4}{5} \div 7 =$

13. $\frac{9}{10} \div 12 =$

14. How many packages of candy, each containing $\frac{3}{4}$ lb., can be filled from 45 lb. of candy?

15. Mrs. Chung bought $8\frac{1}{3}$ yd. of cloth for $50. How much did she pay for each yard?

16. $2\frac{3}{4} \div 3 =$

17. $3\frac{1}{5} \div 4 =$

18. $17\frac{1}{2} \div 10 =$

19. $1\frac{7}{8} \div \frac{15}{32} =$

20. $1\frac{2}{3} \div \frac{3}{4} =$

21. $3\frac{5}{6} \div \frac{2}{3} =$

22. $\frac{5}{16} \div 2\frac{2}{3} =$

23. $\frac{7}{10} \div 5\frac{3}{5} =$

24. $6\frac{3}{4} \div 1\frac{1}{8} =$

25. $4\frac{1}{6} \div 4\frac{3}{8} =$

26. $5\frac{4}{5} \div 2\frac{1}{2} =$

27. $2\frac{11}{32} \div 1\frac{1}{4} =$

28. Ten girls are to share $3\frac{3}{4}$ lb. of cake equally. How much cake does each girl get?

29. How many $4\frac{1}{2}$ in. lengths of ribbon can be cut from a piece $13\frac{1}{2}$ in. long?

30. An athlete participating in the long jump makes three leaps of $22\frac{1}{2}$ ft., 24 ft., and $23\frac{5}{8}$ ft. What is the average length of his three jumps?

Now turn to the end of the chapter to check your answers. Add up all that you had correct.

A Score of	Means That You
27–30	Did very well. You can move to Chapter 10.
24–26	Know the material except for a few points. Read the sections about the ones you missed.
20–23	Need to check carefully on the sections you missed.
0–19	Need to work with this part of the chapter to refresh your memory and improve your skills.

9.7 The Quotient Of Two Fractions

Example

Diana buys $\frac{1}{2}$ lb. of cashew nuts and asks the salesclerk to put the nuts into $\frac{1}{4}$ lb. bags. How many bags does Diana receive from the salesclerk?

Solution

The way to solve this example is to divide $\frac{1}{2}$ by $\frac{1}{4}$, since you are trying to determine the number of quarters there are in one half.

$$\frac{1}{2} \div \frac{1}{4} = ?$$

can be illustrated by the following diagram. Start with $\frac{1}{2}$ by dividing the whole into two equal parts with a solid line. Shade in $\frac{1}{2}$ of the whole.

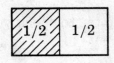

118

Divide the whole into 4 equal parts. You see that there are two quarters in the

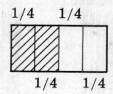

shaded portion equal to $\frac{1}{2}$. Thus,

$$\frac{1}{2} \div \frac{1}{4} = 2$$

While this is a suitable method for finding the answer to simple problems, what will you do when you are faced with more difficult ones? Obviously, an easier method must be found for dealing with all types of division examples involving fractions.

From the previous section on multiplying fractions you know that $\frac{1}{2} \times \frac{4}{1} = 2$. This problem has two elements which are the same as $\frac{1}{2} \div \frac{1}{4} = 2$, namely, the dividend $\frac{1}{2}$ and the quotient 2. The divisor $\frac{1}{4}$ is the *inverse* of $\frac{4}{1}$. The *inverse* of a fraction is the fraction obtained by interchanging its numerator and denominator.

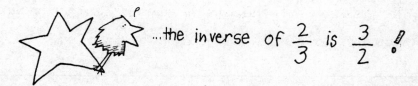

...the inverse of $\frac{2}{3}$ is $\frac{3}{2}$!

The inverse of $\frac{2}{3}$ is $\frac{3}{2}$.

The inverse of $\frac{5}{12}$ is $\frac{12}{5}$.

The inverse of 6 is $\frac{1}{6}$. (Remember: $6 = \frac{6}{1}$).

You notice that the division example $\frac{1}{2} \div \frac{1}{4} = 2$ and the multiplication example $\frac{1}{2} \times \frac{4}{1} = 2$ result in the same solution. This gives you a hint on how to divide fractions. All you have to do is to write the inverse of the divisor, change the operation to multiplication, and then proceed, using the method for multiplying fractions. Look at the following example:

119

Problem	Step 1	Step 2	Step 3
$\frac{1}{2} \div \frac{1}{4} =$	$\frac{1}{2} \times \frac{4}{1} =$	$\frac{1}{\overset{}{2}} \times \frac{\overset{2}{4}}{1} =$ (with $\frac{1}{2}$ marked 1, $\frac{4}{1}$ marked 2)	$\frac{1}{1} \times \frac{2}{1} = \frac{2}{1} = 2$
	Write the inverse of the divisor $\left(\frac{1}{4}\right): \frac{4}{1}$ Change the division operation to multiplication.	Use cancellation: $4 \div 2 = 2$ $2 \div 2 = 1$	Multiply the numerators: $1 \times 2 = 2$ Multiply the denominators: $1 \times 1 = 1$

Thus: $\frac{1}{2} \div \frac{1}{4} = 2$.

"Write the inverse of the divisor" is sometimes called "invert the divisor." That expression may be familiar to you.

And another —

Example

$\frac{5}{6} \div \frac{1}{3} = ?$

Solution

Problem	Step 1	Step 2	Step 3
$\frac{5}{6} \div \frac{1}{3} =$	$\frac{5}{6} \times \frac{3}{1} =$	$\frac{5}{\underset{2}{6}} \times \frac{\overset{1}{3}}{1} =$	$\frac{5}{2} \times \frac{1}{1} = \frac{5}{2} = 2\frac{1}{2}$
	Invert the divisor $\left(\frac{1}{3} \text{ becomes } \frac{3}{1}\right)$ and multiply.	Use cancellation: $6 \div 3 = 2$ $3 \div 3 = 1$	Multiply the numerators: $5 \times 1 = 5$ Multiply the denominators: $2 \times 1 = 2$ Simplify the fraction: $\frac{5}{2} = 2\frac{1}{2}$

Thus: $\frac{5}{6} \div \frac{1}{3} = 2\frac{1}{2}$.

Practice Exercise 102

Divide. Use cancellation when you can and write each answer in its simplest form.

1. $\frac{3}{4} \div \frac{1}{8} =$

2. $\frac{2}{3} \div \frac{1}{4} =$

3. $\frac{1}{2} \div \frac{2}{3} =$

4. $\frac{1}{2} \div \frac{1}{4} =$

5. $\frac{3}{4} \div \frac{1}{2} =$

6. $\frac{2}{3} \div \frac{4}{9} =$

7. $\frac{5}{8} \div \frac{1}{2} =$

8. $\frac{1}{3} \div \frac{1}{4} =$

9. $\frac{7}{8} \div \frac{7}{12} =$

10. $\frac{1}{3} \div \frac{1}{6} =$

11. $\frac{1}{8} \div \frac{1}{2} =$

12. $\frac{2}{3} \div \frac{1}{2} =$

9.8 Dividing Mixed Numbers

The division of mixed numbers is handled in the same way as the multiplication of mixed numbers. You change the mixed number or whole number to a fraction and then proceed as you learned in the previous section.

Remember: *Change the mixed number to a fraction and then invert the divisor and multiply.* Use the example that follows as a model.

Example

A recipe calls for $\frac{1}{4}$ lb. of meat per person. Mrs. Jackson buys 3 lb. of meat. How many people will this feed?

Solution

The solution to this example can be found by dividing 3 by $\frac{1}{4}$, or

$$3 \div \frac{1}{4} = ?$$

Problem	Step 1	Step 2	Step 3
$3 \div \frac{1}{4} =$	$\frac{3}{1} \div \frac{1}{4} =$	$\frac{3}{1} \times \frac{4}{1} =$	$\frac{3}{1} \times \frac{4}{1} = \frac{12}{1} = 12$
	Write the whole number as a fraction: $3 = \frac{3}{1}$	Invert the divisor $\left(\frac{1}{4} \text{ becomes } \frac{4}{1}\right)$ and multiply.	Multiply the numerators: $3 \times 4 = 12$ Multiply the denominators: $1 \times 1 = 1$

Thus: $3 \div \frac{1}{4} = 12$.

And another —

Example

How many lengths of $1\frac{7}{8}$ in. can be cut from an aluminum rod measuring $7\frac{1}{2}$ in.? (Waste from cutting is not considered.)

Solution

The solution to this problem is to divide $7\frac{1}{2}$ by $1\frac{7}{8}$ or

$$7\frac{1}{2} \div 1\frac{7}{8} = ?$$

Picturing this example on a ruler, like that drawn below, you see that there are *four* lengths of $1\frac{7}{8}$ in. in a line $7\frac{1}{2}$ in. in length.

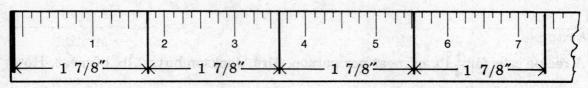

Thus: $7\frac{1}{2} \div 1\frac{7}{8} = 4$.

The shorter arithmetic method requires you to change each of the mixed numbers to improper fractions first and then to continue as you did before.

Problem	Step 1	Step 2	Step 3	Step 4
$7\frac{1}{2} \div 1\frac{7}{8} =$	$\frac{15}{2} \div \frac{15}{8} =$	$\frac{15}{2} \times \frac{8}{15} =$	$\overset{1}{\underset{1}{\cancel{\frac{15}{2}}}} \times \overset{4}{\underset{1}{\cancel{\frac{8}{15}}}} =$	$\frac{1}{1} \times \frac{4}{1} = \frac{4}{1} = 4$
	Change each mixed number to an improper fraction: $7\frac{1}{2} = \frac{15}{2}$ $1\frac{7}{8} = \frac{15}{8}$	Invert the divisor $\left(\frac{15}{8}\right)$ and multiply.	Use cancellation: $15 \div 15 = 1$ $15 \div 15 = 1$ and $8 \div 2 = 4$ $2 \div 2 = 1$	Multiply the numerators: $1 \times 4 = 4$ Multiply the denominators: $1 \times 1 = 1$

Thus: $7\frac{1}{2}$ in. $\div$ $1\frac{7}{8}$ in. $= 4$.

And still another —

Example

$4 \div 6\frac{2}{3} = ?$

Solution

Problem	Step 1	Step 2	Step 3	Step 4
$4 \div 6\frac{2}{3} =$	$\frac{4}{1} \div \frac{20}{3} =$	$\frac{4}{1} \times \frac{3}{20}$	$\overset{1}{\underset{5}{\cancel{\frac{4}{1}}}} \times \frac{3}{\cancel{20}} =$	$\frac{1}{1} \times \frac{3}{5} = \frac{3}{5}$
	Write each as an improper fraction: $4 = \frac{4}{1}$ $6\frac{2}{3} = \frac{20}{3}$	Invert the divisor $\left(\frac{20}{3}\right)$ and multiply.	Use cancellation: $4 \div 4 = 1$ $20 \div 4 = 5$	Multiply: $1 \times 3 = 3$ and $1 \times 5 = 5$

Thus: $4 \div 6\frac{2}{3} = \frac{3}{5}$.

Practice Exercise 103

Divide.

1. $6 \div \frac{2}{3} =$

2. $10 \div \frac{5}{6} =$

3. $4 \div \frac{1}{2} =$

4. $3 \div 1\frac{1}{2} =$

5. $1 \div 2\frac{1}{8} =$

6. $7 \div 2\frac{1}{3} =$

7. $4\frac{2}{5} \div 5\frac{1}{2} =$

8. $1\frac{2}{3} \div 2\frac{3}{5} =$

9. $10\frac{2}{3} \div 5\frac{1}{3} =$

10. $4\frac{5}{6} \div 3 =$

11. $6\frac{2}{3} \div 4 =$

12. $3\frac{7}{12} \div \frac{2}{3} =$

9.9 Word Problems

Reread Section 9.6 of this chapter to be sure you are familiar with the *five* steps necessary to solve a word problem. When you feel sure you know these steps thoroughly, you will be ready to solve these practice exercises.

Practice Exercise 104

1. How many pieces can be cut from a 27 in. aluminum pipe if each piece measures $3\frac{3}{8}$ in.?

2. How many bags of seed weighing $2\frac{1}{2}$ lb. each can be made from a sack weighing 60 lb.?

3. If a board $8\frac{3}{4}$ ft. long is cut into 7 equal parts, how long is each part?

4. Mr. Oak purchased $8\frac{1}{2}$ acres of land and plans to build 17 houses, each on an equal amount of land. How many acres will each plot of land contain?

124

5. Mr. Small has 5 cartons of dress goods which total $182\frac{1}{2}$ lb. Each carton contains an equal weight and no carton can exceed 35 lb. when shipped.

   a. Are the cartons within the limit of 35 lb. each?

   b. How many pounds are they above or below the weight of 35 lb. each?

6. Find the average of the following *three* heights jumped by a pole vaulter in a recent track meet: $12\frac{1}{2}$ ft., $14\frac{7}{8}$ ft., and $14\frac{3}{4}$ ft.

7. If a dress requires $2\frac{9}{16}$ yd. of material, how many dresses could be made if you had 82 yd. of material?

8. A steel bar measures $12\frac{1}{4}$ ft. in length. How many rods $1\frac{3}{4}$ ft. long can be cut from it?

9. A vegetable garden contains 45 sq. ft. If the length of the rectangular garden is $5\frac{5}{8}$ ft., what is its width?

10. A recipe calls for $1\frac{1}{3}$ tablespoons of butter and serves 6 people. If you wish to cook this recipe for yourself, how much butter should you use?

Terms You Should Remember

Quotient   The quantity resulting from dividing one quantity by another.

Invert   To interchange the numerator and denominator of a fraction.

Average   Most often referred to as the *arithmetic mean*.

Review Of Important Ideas

Some of the most important ideas in this part of Chapter 9 were:

 To divide fractions, invert the divisor, change the division sign to a multiplication sign, and proceed as you did when multiplying fractions.

 When dividing mixed numbers, change each mixed number to a fraction, invert the divisor, and proceed as you did when multiplying fractions.

 To find either a length or width of a rectangle, given one of these dimensions and the area, you divide the area by one of the given dimensions. In symbols:

$$L = \frac{A}{W} \text{ or } W = \frac{A}{L}$$

Check What You Have Learned

This test will allow you to see how your skill in computing is progressing. Try to achieve a satisfactory grade.

Posttest 9B

Write your answers below each question.

1. $8 \div \frac{1}{2} =$

2. $24 \div \frac{3}{4} =$

3. $7 \div \frac{4}{5} =$

4. $12 \div 1\frac{1}{5} =$

5. $3 \div 2\frac{5}{6} =$

6. $6 \div 3\frac{3}{4} =$

7. $18 \div 1\frac{5}{16} =$

8. $\frac{2}{3} \div \frac{3}{4} =$

9. $\frac{3}{4} \div \frac{15}{16} =$

10. $\frac{5}{9} \div \frac{5}{27} =$

11. $\frac{5}{6} \div \frac{7}{18} =$

12. $\frac{2}{5} \div 9 =$

13. $\frac{9}{10} \div 15 =$

14. How many cans of nuts, each containing $\frac{5}{6}$ lb., can be filled from 40 lb. of nuts?

15. Jean bought $7\frac{1}{2}$ yd. of material for $60. How much did she pay for one yard?

16. $3\frac{2}{5} \div 4 =$ 17. $2\frac{1}{4} \div 3 =$ 18. $12\frac{1}{2} \div 20 =$ 19. $1\frac{5}{8} \div \frac{13}{16} =$

20. $1\frac{3}{4} \div \frac{4}{5} =$ 21. $3\frac{5}{8} \div \frac{3}{4} =$ 22. $\frac{7}{20} \div 3\frac{1}{3} =$ 23. $\frac{3}{10} \div 5\frac{2}{5} =$

24. $6\frac{2}{3} \div 1\frac{1}{9} =$ 25. $3\frac{3}{4} \div 4\frac{1}{6} =$ 26. $4\frac{1}{5} \div 1\frac{2}{3} =$ 27. $2\frac{17}{32} \div 2\frac{1}{4} =$

28. Nine girls want to share $5\frac{1}{4}$ pounds of candy equally. How much candy should each girl get?

29. How many $3\frac{1}{2}$ in. pieces of ribbon can be cut from a piece $17\frac{1}{2}$ in. long?

30. Three track stars run a race in times of $39\frac{1}{2}$ sec., $38\frac{3}{4}$ sec., and $39\frac{1}{8}$ sec. What is the average time for this event?

ANSWERS AND EXPLANATIONS
TO POSTTEST 9B

---

1. $8 \div \frac{1}{2} = \frac{8}{1} \times \frac{2}{1} = 16$

2. $24 \div \frac{3}{4} = \frac{\overset{8}{\cancel{24}}}{1} \times \frac{4}{\cancel{3}} = 32$
   $\phantom{24 \div \frac{3}{4} = }{\scriptstyle 1}$

3. $7 \div \frac{4}{5} = \frac{7}{1} \times \frac{5}{4} = 8\frac{3}{4}$

4. $12 \div 1\frac{1}{5} = \frac{12}{1} \div \frac{6}{5} = \frac{\overset{2}{\cancel{12}}}{1} \times \frac{5}{\cancel{6}} = 10$
   $\phantom{12 \div 1\frac{1}{5} = \frac{12}{1} \div \frac{6}{5} = }{\scriptstyle 1}$

5. $3 \div 2\frac{5}{6} = \frac{3}{1} \div \frac{17}{6} = \frac{3}{1} \times \frac{6}{17} = 1\frac{1}{17}$

6. $6 \div 3\frac{3}{4} = \frac{6}{1} \div \frac{15}{4} = \frac{\overset{2}{\cancel{6}}}{1} \times \frac{4}{\cancel{15}} = 1\frac{3}{5}$
   $\phantom{6 \div 3\frac{3}{4} = \frac{6}{1} \div \frac{15}{4} = }{\scriptstyle 5}$

7. $18 \div 1\frac{5}{16} = \frac{18}{1} \div \frac{21}{16} = \frac{\overset{6}{\cancel{18}}}{1} \times \frac{16}{\underset{7}{\cancel{21}}} = 13\frac{5}{7}$

8. $\frac{2}{3} \div \frac{3}{4} = \frac{2}{3} \times \frac{4}{3} = \frac{8}{9}$

9. $\frac{3}{4} \div \frac{15}{16} = \frac{\overset{1}{\cancel{3}}}{\underset{1}{\cancel{4}}} \times \frac{\overset{4}{\cancel{16}}}{\underset{5}{\cancel{15}}} = \frac{4}{5}$

10. $\frac{5}{9} \div \frac{5}{27} = \frac{\overset{1}{\cancel{5}}}{\cancel{9}} \times \frac{\overset{3}{\cancel{27}}}{\underset{1}{\cancel{5}}} = 3$

11. $\frac{5}{6} \div \frac{7}{18} = \frac{5}{\underset{1}{\cancel{6}}} \times \frac{\overset{3}{\cancel{18}}}{7} = 2\frac{1}{7}$

12. $\frac{2}{5} \div 9 = \frac{2}{5} \div \frac{9}{1} = \frac{2}{5} \times \frac{1}{9} = \frac{2}{45}$

13. $\frac{9}{10} \div \frac{15}{1} = \frac{\overset{3}{\cancel{9}}}{10} \times \frac{1}{\underset{5}{\cancel{15}}} = \frac{3}{50}$

14. $40 \div \frac{5}{6} = \frac{\overset{8}{\cancel{40}}}{1} \times \frac{6}{\underset{1}{\cancel{5}}} = 48 \text{ cans}$

15. $\$60 \div 7\frac{1}{2} = \frac{60}{1} \div \frac{15}{2} = \frac{\overset{4}{\cancel{60}}}{1} \times \frac{2}{\underset{1}{\cancel{15}}} = \$8$

16. $3\frac{2}{5} \div 4 = \frac{17}{5} \div \frac{4}{1} = \frac{17}{5} \times \frac{1}{4} = \frac{17}{20}$

17. $2\frac{1}{4} \div 3 = \frac{9}{4} \div \frac{3}{1} = \frac{\overset{3}{\cancel{9}}}{4} \times \frac{1}{\underset{1}{\cancel{3}}} = \frac{3}{4}$

18. $12\frac{1}{2} \div 20 = \frac{25}{2} \div \frac{20}{1} = \frac{\overset{5}{\cancel{25}}}{2} \times \frac{1}{\underset{4}{\cancel{20}}} = \frac{5}{8}$

19. $1\frac{5}{8} \div \frac{13}{16} = \frac{13}{8} \div \frac{13}{16} = \frac{\overset{1}{\cancel{13}}}{\underset{1}{\cancel{8}}} \times \frac{\overset{2}{\cancel{16}}}{\underset{1}{\cancel{13}}} = 2$

20. $1\frac{3}{4} \div \frac{4}{5} = \frac{7}{4} \div \frac{4}{5} = \frac{7}{4} \times \frac{5}{4} = 2\frac{3}{16}$

21. $3\frac{5}{8} \div \frac{3}{4} = \frac{29}{8} \div \frac{3}{4} = \frac{29}{\underset{2}{\cancel{8}}} \times \frac{\cancel{4}}{3} = 4\frac{5}{6}$

22. $\frac{7}{20} \div 3\frac{1}{3} = \frac{7}{20} \div \frac{10}{3} = \frac{7}{20} \times \frac{3}{10} = \frac{21}{200}$

23. $\frac{3}{10} \div 5\frac{2}{5} = \frac{3}{10} \div \frac{27}{5} = \frac{\overset{1}{\cancel{3}}}{\underset{2}{\cancel{10}}} \times \frac{\overset{1}{\cancel{5}}}{\underset{9}{\cancel{27}}} = \frac{1}{18}$

24. $6\frac{2}{3} \div 1\frac{1}{9} = \frac{20}{3} \div \frac{10}{9} = \frac{\overset{2}{\cancel{20}}}{\underset{1}{\cancel{3}}} \times \frac{\overset{3}{\cancel{9}}}{\underset{1}{\cancel{10}}} = 6$

25. $3\frac{3}{4} \div 4\frac{1}{6} = \frac{15}{4} \div \frac{25}{6} = \frac{\overset{3}{\cancel{15}}}{\underset{2}{\cancel{4}}} \times \frac{\overset{3}{\cancel{6}}}{\underset{5}{\cancel{25}}} = \frac{9}{10}$

26. $4\frac{1}{5} \div 1\frac{2}{3} = \frac{21}{5} \div \frac{5}{3} = \frac{21}{5} \times \frac{3}{5} = 2\frac{13}{25}$

27. $2\frac{17}{32} \div 2\frac{1}{4} = \frac{81}{32} \div \frac{9}{4} = \frac{\overset{9}{\cancel{81}}}{\underset{8}{\cancel{32}}} \times \frac{\overset{1}{\cancel{4}}}{\underset{1}{\cancel{9}}} = 1\frac{1}{8}$

28. $5\frac{1}{4} \div 9 = \frac{21}{4} \div \frac{9}{1} = \frac{\overset{7}{\cancel{21}}}{4} \times \frac{1}{\underset{3}{\cancel{9}}} = \frac{7}{12} \text{ lb.}$

29. $17\frac{1}{2} \div 3\frac{1}{2} = \frac{35}{2} \div \frac{7}{2} = \frac{\overset{5}{\cancel{35}}}{\underset{1}{\cancel{2}}} \times \frac{\overset{1}{\cancel{2}}}{\underset{1}{\cancel{7}}} = 5$

30. $$\frac{39\frac{1}{2} + 38\frac{3}{4} + 39\frac{1}{8}}{3} = \frac{117\frac{3}{8}}{3} = 117\frac{3}{8} \div 3 = \frac{\overset{313}{\cancel{939}}}{8} \times \frac{1}{\underset{1}{\cancel{3}}} = 39\frac{1}{8} \text{ sec.}$$

A Score of	Means That You
27–30	Did very well. You can move to Chapter 10.
24–26	Know this material except for a few points. Reread the sections about the ones you missed.
20–23	Need to check carefully on the sections you missed.
0–19	Need to review this part of the chapter again to refresh your memory and improve your skills.

Questions	Are Covered in Section
8–11	9.7
1–7, 12, 13, 16–27	9.8
14, 15, 28–30	9.9

ANSWERS FOR CHAPTER 9 — B

PRETEST 9B

1. 27
2. 25
3. $13\frac{1}{3}$
4. 8
5. $1\frac{11}{21}$
6. $\frac{8}{9}$
7. $9\frac{3}{5}$
8. $1\frac{1}{8}$
9. $\frac{2}{3}$
10. 4
11. $1\frac{3}{5}$
12. $\frac{4}{35}$
13. $\frac{3}{40}$
14. 60 pkges.
15. $6
16. $\frac{11}{12}$
17. $\frac{4}{5}$
18. $1\frac{3}{4}$
19. 4
20. $2\frac{2}{9}$
21. $5\frac{3}{4}$
22. $\frac{15}{128}$
23. $\frac{1}{8}$
24. 6
25. $\frac{20}{21}$
26. $2\frac{8}{25}$
27. $1\frac{7}{8}$
28. $\frac{3}{8}$ lbs.
29. 3 lengths
30. $23\frac{3}{8}$ ft.

PRACTICE EXERCISE 102

1. 6
2. $2\frac{2}{3}$
3. $\frac{3}{4}$
4. 2

5. $1\frac{1}{2}$ 6. $1\frac{1}{2}$ 7. $1\frac{1}{4}$ 8. $1\frac{1}{3}$

9. $1\frac{1}{2}$ 10. 2 11. $\frac{1}{4}$ 12. $1\frac{1}{3}$

PRACTICE EXERCISE 103

1. 9 2. 12 3. 8 4. 2

5. $\frac{8}{17}$ 6. 3 7. $\frac{4}{5}$ 8. $\frac{25}{39}$

9. 2 10. $1\frac{11}{18}$ 11. $1\frac{2}{3}$ 12. $5\frac{3}{8}$

PRACTICE EXERCISE 104

1. $27 \div 3\frac{3}{8} = \frac{27}{1} \div \frac{27}{8} = \frac{27}{1} \times \frac{8}{27} = 8$ pieces

2. $60 \div 2\frac{1}{2} = \frac{60}{1} \div \frac{5}{2} = \frac{60}{1} \times \frac{2}{5} = 24$ bags

3. $8\frac{3}{4} \div 7 = \frac{35}{4} \div \frac{7}{1} = \frac{35}{4} \times \frac{1}{7} = \frac{5}{4} = 1\frac{1}{4}$ ft.

4. $8\frac{1}{2} \div 17 = \frac{17}{2} \div \frac{17}{1} = \frac{17}{2} \times \frac{1}{17} = \frac{1}{2}$ acre

5. $182\frac{1}{2} \div 5 = \frac{365}{2} \div \frac{5}{1} = \frac{365}{2} \times \frac{1}{5} = \frac{73}{2} = 36\frac{1}{2}$ lb.   a. No   b. $1\frac{1}{2}$ lb. above weight

6. $\dfrac{12\frac{1}{2} + 14\frac{7}{8} + 14\frac{3}{4}}{3} = \dfrac{42\frac{1}{8}}{3} = 42\frac{1}{8} \div 3 = \frac{337}{8} \times \frac{1}{3} = \frac{337}{24} = 14\frac{1}{24}$ ft.

7. $82 \div 2\frac{9}{16} = \frac{82}{1} \div \frac{41}{16} = \frac{82}{1} \times \frac{16}{41} = 32$ dresses

8. $12\frac{1}{4} \div 1\frac{3}{4} = \frac{49}{4} \div \frac{7}{4} = \frac{49}{4} \times \frac{4}{7} = 7$ rods

9. $W = \frac{A}{L}$ $W = \frac{45}{5\frac{5}{8}} = \frac{45}{1} \div \frac{45}{8} = \frac{45}{1} \times \frac{8}{45} = 8$ ft.

10. $1\frac{1}{3} \div 6 = \frac{4}{3} \div \frac{6}{1} = \frac{4}{3} \times \frac{1}{6} = \frac{2}{9}$ tablespoons

130

Hold It!

Here is your chance to see how your skills in computing are returning to you. This exercise reviews some of the work covered so far. Try hard for a perfect score.

1. $3{,}432 + 732 + 43{,}970 + 9 =$

2. Find the difference of 5,000 and 2,739.

3. Add: $3\frac{7}{8} + 4\frac{1}{6}$

4. $37 \overline{)\, 3774}$

5. $\begin{array}{r} 673 \\ \times\ 45 \\ \hline \end{array}$

6. $\begin{array}{r} 193\frac{1}{2} \\ -146\frac{3}{8} \\ \hline \end{array}$

7. $6 \times 1\frac{1}{4} =$

8. $3\frac{5}{8} \div 1\frac{1}{16} =$

9. If $3\frac{5}{8}$ yd. of material are needed to make a dress, how many yards are needed to make 96 dresses?

10. If $13\frac{9}{16}$ ft. are cut from a 16 ft. board, how many feet still remain?

ANSWERS TO "HOLD IT!"

1.	48,143	2.	2,261	3.	$8\frac{1}{24}$	4.	102
5.	30,285	6.	$47\frac{1}{8}$	7.	$7\frac{1}{2}$	8.	$3\frac{7}{17}$
9.	348 yd.	10.	$2\frac{7}{16}$ ft.				

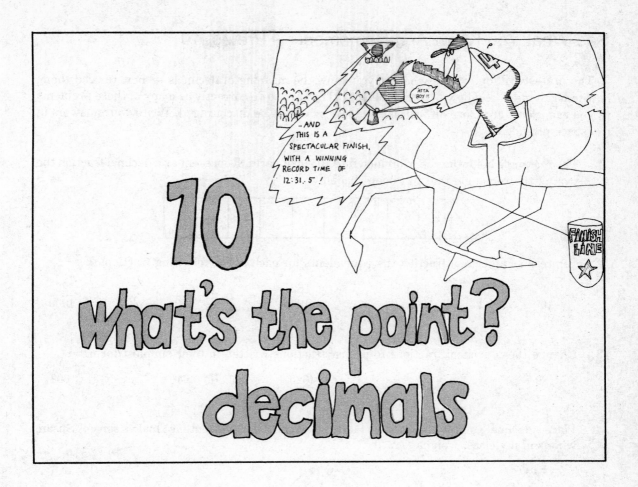

10
what's the point?
decimals

Decimal fractions play a major role in our daily lives. The most use of decimals by the average person is in relation to money. Machine shop operators, parts inspectors, and other skilled persons concerned with more precise measurements than one can read from a ruler need to know decimals very well.

You have certainly seen many instances where decimals have an important role. Baseball players have batting averages of .283; pitchers have earned run averages of 3.54; track races are timed in 9.1 sec.; and distances between cities are written as 38.7 mi.; all are examples of decimal fractions or decimal numerals.

You probably know other examples of the same type. Decimals are no more difficult to use than the fractions you have completed. As a matter of fact, decimals may be easier for you to learn now that you understand fractions.

These next four chapters deal with this type of number, its meaning, its operations, and its relationship to fractions.

See What You Know And Remember — Pretest 10

The pretest will determine whether you know the meaning of decimals — how to read them, how to write them, and how to round them off. Try hard to do correctly as many of these problems as you can. Some are more difficult than others so do not be discouraged. Write your answers in the space provided.

1. The rectangle below has been divided into 10 equal parts. Represent as a decimal fraction the part of the rectangle that has been shaded.

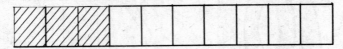

2. Express as a decimal fraction the equivalents for each of the following fractions.

   a. $\dfrac{6}{10}$ b. $\dfrac{48}{100}$ c. $\dfrac{162}{1000}$ d. $\dfrac{5}{100}$ e. $\dfrac{22}{1000}$

3. Change these decimal fractions to proper fractions written in their simplest form.

   a. .39   b. .8   c. .625   d. .04   e. .002

4. Place a decimal point in the number 1125 so that the following sentence makes sense: "Susan, who is in my class, weighs *1125* lb."

···OH DEAR!

5. Mr. Brown earns $100 a week. He spends $86 and saves the rest. Express, as a decimal fraction, the part of his salary he saves.

6. Eight-thousandths is written as

   a. 8.000   b. .008   c. .0008   d. .8000

7. a. Write, in words, the meaning of the decimal fraction: .2.

   b. Write, in words, the meaning of the decimal numeral: 3.07.

134

8. Write in figures, the number represented by:

    a. forty-four hundredths

    b. three and two-tenths

    c. three thousand four and sixty-three thousandths

9. Place one of the symbols >, <, or = within the parentheses to make each of the following statements true:

    a. .8 (   ) .4   b. .50 (   ) .5   c. .070 (   ) .70

    d. .1 (   ) .0925   e. 2.0 (   ) 2.009

10. Arrange each decimal in order of size with the smallest number first.

    a. .6, .06, .65

    b. 7.37, 7.037, 7.307

    c. 2.55, 2.505, 2.5

11. Round off each of the following numbers to the indicated place.

    a. .87 to the nearest tenth

    b. 11.434 to the nearest hundredth

    c. 3.1416 to the nearest thousandth

Now turn to the end of the chapter to check your answers. Add up all that you had correct. Count by the number of separate answers, not by the number of questions. In this pretest there were 11 questions, but 30 separate answers.

A Score of	Means That You
27–30	Did very well. You can move to Chapter 11.
24–26	Know this material except for a few points. Read the sections about the ones you missed.
20–23	Need to check carefully on the sections you missed.
0–19	Need to work with this chapter to refresh your memory and improve your skills.

10.1 Introducing Decimals

You have previously learned that each digit in a numeral has two values —*face* and *place* values. In the numeral 234, the digits 2, 3, and 4 have *face* values equivalent to 2, 3, and 4 units respectively. But because of their position in the number, they have a second value. Look at the illustration which follows.

In the numeral 234,

2 represents 2 hundreds
3 represents 3 tens
4 represents 4 ones

Each digit has a place value which is always 10 times the place value of the digit immediately to its right and one-tenth as much as the place value of the digit immediately to its left. In tabular form it looks like this:

	TEN-THOUSANDS	THOUSANDS	HUNDREDS	TENS	ONES
etc. ⟵ ⟵					
	$10 \times 10 \times 10 \times 10 \times 1 =$ 10,000	$10 \times 10 \times 10 \times 1 =$ 1,000	$10 \times 10 \times 1 = 100$	$10 \times 1 = 10$	1
			2	3	4

136

Clearly, 234 belongs in the three columns to the right, as shown below the table, marked with an ———→ . The 2 is in the hundreds column, the 3 is in the tens column, and the 4 is in the ones column.

You know that there are other place values farther to the left than are indicated on the chart. Hundred-thousands, millions, etc., are all columns to the left of those listed. Each of these columns has a place value that is 10 times the column to its right. Thousands are 10 times the hundreds column, hundreds are 10 times the tens column, and so on.

Let us consider the relationship of any column to the one on its right. The column to the right is one-tenth or $\frac{1}{10}$ of the column on its left. This is illustrated in the following table:

10,000	1,000	100	10	1	$\frac{1}{10}$	$\frac{1}{100}$	$\frac{1}{1,000}$	$\frac{1}{10,000}$
		2	3	4 .	8	7	5	
...TEN THOUSANDS	THOUSANDS	HUNDREDS	TENS	ONES .	TENTHS	HUNDREDTHS	THOUSANDTHS	TEN THOUSANDTHS...

Each place has a value one-tenth of that to the left. Each place has a value ten times the place to the right.

The place value of the column to the right of the ones column is one-tenth of its place value and is called the *tenths* column. This column represents fractional values and each column to its right also represents fractional values.

Looking at it again, you see that the place value of the column to the right of the ones column must be multiplied by 10 and the product must be the place value of the ones column, or 1. Clearly, the problem becomes

$$10 \times ? = 1$$

The answer is $\frac{1}{10}$, since $10 \times \frac{1}{10} = 1$.

Thus, in the numeral 234.875, at the bottom of the table, eight-tenths $= .8 = \frac{8}{10}$. This is a *one-place* decimal. It is written as .8 and read as 8 tenths.

Practice Exercise 105

Write each of the following in decimal form.

1. $\frac{6}{10}$ 2. $\frac{3}{10}$ 3. $\frac{1}{10}$ 4. $\frac{9}{10}$ 5. $\frac{7}{10}$

Write each of the following in fractional form.

6. .5   7. .2   8. .8   9. .1   10. .4

Moving to the next column to the right, you have

$$10 \times ? = \frac{1}{10}$$

The solution to this example is $\frac{1}{100}$, and the column to the right of the tenths column is called the *hundredths* column. Other columns are found the same way and are called thousandths, ten-thousandths, hundred-thousandths, etc. You may continue indefinitely. You should also notice that as you move to the right values are shrinking. One-thousandth is much less than one-tenth:

$$\frac{1}{1,000} < \frac{1}{10}$$

The *decimal point* (.) separates the whole-number portion of the decimal numeral from the fractional portion. Thus, in the decimal numeral 234.875, illustrated at the foot of the table, 234 is the whole-number part of the numeral and 875 is the fractional part. This is called a *mixed decimal*.

$$.8 = \frac{8}{10}$$

$$.87 = \frac{87}{100}$$

$$.875 = \frac{875}{1,000}$$

Example

During the first two weeks in September $\frac{32}{100}$ in. of rain fell. Write this as a decimal fraction.

Solution

$$\frac{32}{100} = .32$$

This is a two-place decimal. It is written as .32 and read as 32 hundredths.

Practice Exercise 106

Write each of the following in decimal form.

1. $\frac{37}{100}$　　　　2. $\frac{463}{1,000}$　　　　3. $\frac{99}{100}$　　　　4. $\frac{4}{100}$

5. $\frac{36}{1,000}$　　　　6. $\frac{7834}{10,000}$　　　　7. $\frac{834}{10,000}$　　　　8. $\frac{1}{1,000}$

Write each of the following in fraction form.

9. .53　　　　10. .25　　　　11. .08　　　　12. .032

13. .205   14. .1034   15. .0078   16. .005

10.2   Reading And Writing Decimals

In the table showing our decimal system (p. 137), you notice that numbers to the right of the decimal point are common fractions whose denominators are powers of 10, such as 10, 100, 1,000, etc.

A *one-place* decimal is a common fraction whose denominator is 10, like $\frac{6}{10}$. It is written as .6 and read as 6 tenths.

A *two-place* decimal is a common fraction whose denominator is 100, like $\frac{6}{100}$. It is written as .06 and read as 6 hundredths.

When a decimal has *three places* to the right of the decimal point, it is read as thousandths. Therefore, .006 is read as 6 thousandths.

Notice that the zeros in the number are quite important since they help place the digit or digits in their proper place. Without them, each number in the previous illustrations would look like .6 and each would be read as 6 tenths.

Suppose zero is to the right of a decimal fraction. What is the value of .60? The value of .60 is the same as .6, showing us that the *zero is meaningless when it is to the right of the decimal point and to the right of the decimal fraction.*

Example

Read and write the decimal fraction .25.

Solution

1.   The numeral .25 is a *two-place* decimal.
2.   Read it as: 25 hundredths.
3.   Write it as: twenty-five hundredths.

And another —

Example

In the men's 60-yd. dash, the time of the winner was 5.8 sec. Read and write the decimal.

Solution

1.   This is a *mixed decimal* containing a whole number and a one-place decimal fraction.
2.   Read it as: 5 *and* 8 tenths (*and* represents the decimal point).
3.   Write it as: five and eight-tenths.

And another —

Example

Some instruments measure as accurately as one-thousandth of an inch. Write this as a decimal fraction.

Solution

Since thousandths is the third place in the decimal, you place the 1 in that place and fill the missing spaces with zeros. Thus: one-thousandth = .001.

Practice Exercise 107

Write in words the meaning of each decimal.

1. .9

2. .27

3. .05

4. 6.1

5. 3.141

6. .0051

7. $.05\frac{1}{4}$

8. 7.02

Write each of the following as a decimal numeral.

9. 6 and 9 hundredths

10. 3 thousandths

11. 3 and 5 tenths

12. 312 and 4 hundredths

13. $4\frac{3}{4}$ hundredths

14. 93 thousandths

15. 7 and 5 ten-thousandths

16. 40 and 43 hundredths

Write each of the following as a decimal numeral.

17. nine-tenths

18. twenty-five hundredths

19. two hundred seven thousandths

20. three and fourteen thousandths

21. thirty-seven and three hundredths

22. nine hundred and nine thousandths

10.3   Which Fraction Equals Which Decimal?

Decimal numerals which terminate or end can be written as a fraction. To write a decimal as a fraction is as easy as reading a decimal.

Example

Write the decimal .8 as a fraction in its simplest form.

Solution

Problem	Step 1	Step 2
.8 =	8 tenths =	$\frac{8}{10} = \frac{4}{5}$
	Read the decimal.	Write the decimal as a fraction. Simplify the fraction.

Thus: $.8 = \frac{4}{5}$.

And another —

Example

Mr. Washington reads a dimension on a blueprint as .875 in. What measurement on a standard ruler is this equivalent to?

Solution

Problem	Step 1	Step 2
.875 =	875 thousandths =	$\frac{875}{1,000} = \frac{7}{8}$ in.
	Read the decimal.	Write the decimal as a fraction. Simplify the fraction.

Thus: $.875$ in. $= \frac{7}{8}$ in.

Practice Exercise 108

Express each of these examples as a fraction in its simplest form (lowest terms).

1. .5

2. .25

3. .625

4. .75

5. .34

6. .125

7. .9

8. .12

9. .08

Express each of these examples as a mixed number in its simplest form.

10. 3.7

11. 5.75

12. 12.1

13. 6.375

14. 1.875

15. 123.5

10.4 Comparing Decimals

Example

When purchasing a sheet of aluminum, Mr. Green had to choose between two different thicknesses. One sheet measured .09 in. while the other was .11 in. Mr. Green desired the thicker of the two sheets of aluminum. Which did he choose?

Solution

In this problem you must compare the sizes of the two decimals .09 and .11. If you change each decimal to a fraction you will see that it is easy to answer the question. Look below:

$$.09 = \frac{9}{100}$$
$$.11 = \frac{11}{100}$$

Since $\frac{11}{100} > \frac{9}{100}$, it follows that .11 > .09.

Comparing decimals can be done in a simpler way as shown in the following problem.

Example

Compare the decimals .5 and .505 to determine the smaller of the two.

Solution

In the previous problem, the two decimals, .09 and .11, were changed to fractions and then compared. This was easily done since both decimals had denominators of 100. In this example, .5 has a denominator of 10 and .505 has a denominator of 1,000. This can be solved in this way:

Problem	Step 1	Step 2
Compare .5 and .505.	$.5 = .500 = \dfrac{500}{1,000}$ $.505 = \dfrac{505}{1,000}$	$\dfrac{500}{1,000} < \dfrac{505}{1,000}$ or .500 < .505 or .5 < .505
	.505 is a three-place decimal. Change .5 to a three-place decimal by adding two zeros to the right of the decimal: .5 = .500 Change both decimals to fractions.	Compare the fractions.

Thus: .5 < .505.

Example

Arrange these numbers in order of size with the smallest first:

.2, .22, .022

Solution

1. .2 is a *one*-place decimal
2. .22 is a *two*-place decimal
3. .022 is a *three*-place decimal
4. The number .022 has the largest number of decimal places; therefore each of the other decimals will have to become three-place decimals by adding zeros.

Problem	Step 1	Step 2
Compare .2, .22, .022.	$.2 = .200 = \dfrac{200}{1,000}$ $.22 = .220 = \dfrac{220}{1,000}$ $.022 = .022 = \dfrac{22}{1,000}$	$\dfrac{20}{1,000} < \dfrac{200}{1,000} < \dfrac{220}{1,000}$ or $.022 < .2 < .22$
	Since .022 is a *three*-place decimal, change both .2 and .22 to three-place decimals also. To .2 add two zeros: .2 = .200. To .22 add one zero: .22 = .220. Convert each decimal to an equivalent common fraction; all will have the same denominator.	Compare the fractions. Write the equivalent decimals.

Thus: $.022 < .2 < .22$.

Practice Exercise 109

Change each of these decimals to *equivalent* ones expressed as thousandths.

1. .43   2. .04   3. .1

Choose the decimals which have the same value.

4. .56, .506, .560   5. .05, .055, .05000, .050

Select the larger of the two numbers.

6. .82 and .8   7. .95 and .92   8. .71 and .69

9. .004 and .04   10. .1 and .01   11. 4.01 and 3.9

Arrange the numbers in order of size, with the smallest one first.

12. .077, .07, .7   13. 3.01, 3.001, 3.1

10.5 Rounding Off Decimals

Example

Ms. Black works as a payroll secretary and must figure the amount of money to be withheld for Social Security taxes for each employee. While computing the tax for an employee, she figured out that she had to deduct $7.156 from his salary. How much money was deducted?

Solution

$7.156 is a sum of money greater than $7.15 and less than $7.16 or

$$\$7.15 < \$7.156 < \$7.16$$

Ms. Black realized that the first place past the decimal point represents tenths and, in this case, tenths of a dollar, or dimes. The second place represents hundredths of a dollar, or pennies. The third place represents thousandths of a dollar and is called mills, but there is no coin which is equivalent to a mill. Ms. Black had a decision to make — deduct $7.15 or $7.16. The process involved is called *rounding off a decimal*.

Since the digit in the thousandths place is 5 or greater than 5, the digit in the hundredths place is raised to 6. Thus, $7.16 was deducted from the employee's salary.

This illustrates *rounding off a decimal* to the *nearest hundredth*. All of the digits to the left of that place remain the same, the 7 and the 1. The digit in the hundredths place is changed upward to 6 because the digit in the thousandths place is 5 or greater than 5. Look at this illustration:

Problem	Step 1	Step 2
Round off $7.156 to the nearest cent (hundredths).	$7.156	$7.16
	Underline the digit in the hundredths place (5).	All the digits to the left of 5 remain the same: $7.1_. Since the digit in the thousandths place (6) is 5 or greater than 5, the digit in the hundredths place is increased by one.

146

Thus: $7.156 = $7.16.

And another —

Example

Suppose you had the decimal $7.152 and you wished to round it off to the nearest cent.

Solution

Problem	Step 1	Step 2
Round off $7.152 to the nearest cent.	$7.15<u>2</u>	$7.15
	Underline the digit in the hundredths place (5).	All the digits to the left of 5 remain the same: $7.1_ Since the digit in the thousandths place (2) is less than 5, the digit in the hundredths place remains the same.

Thus: $7.152 = $7.15.

Practice Exercise 110

Round off each of the following decimals to the nearest tenth.

1. .34
2. 2.381
3. 3.07
4. .67

Round off each of the following decimals to the nearest hundredth.

5. 4.345
6. .768
7. 2.796
8. .871

Round off each of the following decimals to the nearest thousandth.

9. 3.1416
10. .6784
11. 5.5555

Review Of Important Ideas

Some of the most important ideas in Chapter 10 were:

To read a decimal, a mixed decimal, or a decimal fraction —
1. Read the whole number portion.
2. Read the decimal point as *and*.
3. Read the number to the right of the decimal point and then name the position of the last digit on the right (.04 is read as 4 hundredths since the 4 is in the hundredths place).

To write a decimal numeral —
1. Write the whole number portion first.
2. Place the decimal point.
3. Write the decimal fraction, placing the last digit in the number in the place corresponding to the position stated (to write 3 thousandths, place the 3 in the third place in the decimal and fill in with zeros: .003).

To compare decimals —
1. Determine which of the numbers you are comparing has the greater number of decimal places.
2. Change each of the decimals to an equivalent one by adding zeros to the right of the number so that each decimal has the same number of places.
3. Change each decimal to an equivalent common fraction.
4. Compare the resulting numbers.

To round off decimals —
1. Underline the digit in the decimal place to which you wish to round off the number.

2. Write down all the digits to the left of the underlined digit as part of the answer.
3. Check the digit to the right of the underlined digit:
   a. If this digit is less than 5, the underlined digit remains the same.
   b. If this digit is 5 or greater than 5, the underlined digit is increased by one.

Check What You Have Learned

Good luck on the posttest. See if you can achieve a satisfactory grade.

Posttest 10

Solve these problems, reducing fractions to their lowest terms. Write your answers in the space provided.

1. This rectangle has been divided into 10 equal parts. Represent as a decimal fraction the part of the rectangle that has been shaded.

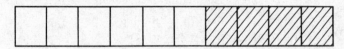

2. Express as a decimal fraction the equivalents for each of the following fractions:

   a. $\frac{5}{10}$ b. $\frac{34}{100}$ c. $\frac{612}{1,000}$ d. $\frac{4}{100}$ e. $\frac{87}{1,000}$

3. Change these decimal fractions to proper fractions written in their simplest form.

   a. .93   b. .6   c. .875   d. .06   e. .004

4. Place a decimal point in the number 2146 so that the following statement makes sense. "John, who is in my adult education class, weighs *2146* lb."

5. Mr. Greene earns $1,000 a month. He spends $960 and saves the rest. Express, as a decimal fraction, the part of his salary he saves.

6. Six-thousandths is written as

   a. 6.000   b. .006   c. .06   d. .6000

149

7. a. Write, in words, the meaning of the decimal fraction: .4.

   b. Write, in words, the meaning of the decimal numeral: 7.03.

8. Write, in figures, the number represented by:

   a. thirty-six hundredths

   b. two and three-tenths

   c. four thousand three and thirty-six thousandths

9. Place one of the symbols >, <, or = within the parentheses to make each of the following statements true.

   a. .7 (　　　) .5　　　b. .60 (　　　) .6　　　c. .090 (　　　) .90
   d. .4 (　　　) .3925　　e. 4.0 (　　　) 4.001

10. Arrange the decimals in order of size with the smallest number first.

    a. .5, .05, .56

    b. 8.34, 8.034, 8.304

    c. 5.22, 5.202, 5.2

11. Round off each of these numbers to the indicated place.

    a. .78 to the nearest tenth

    b. 14.212 to the nearest hundredth

    c. 4.1518 to the nearest thousandth

ANSWERS AND EXPLANATIONS
TO POSTTEST 10

1. $\frac{4}{10} = .4$

2. a. .5　　　b. .34　　　c. .612　　　d. .04　　　e. .087

3. a. $\frac{93}{100}$　　b. $\frac{6}{10} = \frac{3}{5}$　　c. $\frac{875}{1,000} = \frac{7}{8}$　　d. $\frac{6}{100} = \frac{3}{50}$　　e. $\frac{4}{1,000} = \frac{1}{250}$

4. 214.6

5. $1,000 $\frac{40}{1,000} = .040$ or $.04$
    $-$ 960
       ─────
       $40

6. (b) $\frac{6}{1,000} = .006$

7. a. four-tenths   b. seven and three-hundredths

8. a. .36   b. 2.3   c. 4,003.036

9. a. .7 > .5   b. .60 = .6   c. .090 < .90   d. .4 > .3925   e. 4.0 < 4.001

10. a. .05 < .50 < .56 or .05 < .5 < .56
     b. 8.034 < 8.304 < 8.340 or 8.034 < 8.304 < 8.34
     c. 5.200 < 5.202 < 5.220 or 5.2 < 5.202 < 5.22

11. a. .78 = .8   b. 14.212 = 14.21   c. 4.1518 = 4.152

In counting up your answers, remember that there were 30 separate answers in this test.

A Score of	Means That You
27–30	Did very well. You can move to Chapter 11.
24–26	Know this material except for a few points. Reread the sections about the ones you missed.
20–23	Need to check carefully on the sections you missed.
0–19	Need to review the chapter again to refresh your memory and improve your skills.

Questions	Are Covered in Section
1–3, 5	10.1
4, 6	10.2
7, 8	10.3
9, 10	10.4
11	10.5

X   ANSWERS FOR CHAPTER 10

PRETEST 10

1. .3

2. a. .6   b. .48   c. .162   d. .05   e. .022

3. a. $\frac{39}{100}$ b. $\frac{8}{10} = \frac{4}{5}$ c. $\frac{625}{1,000} = \frac{5}{8}$ d. $\frac{4}{100} = \frac{1}{25}$ e. $\frac{2}{1,000} = \frac{1}{500}$

4. 112.5 lb.

5. $100
   − 86
   ────
   $14

$\frac{14}{100} = .14$ savings

6. (b) .008

7. a. two-tenths   b. three and seven hundredths

8. a. .44   b. 3.2

c. 3,004.063

9. a. >   b. =

c. <   d. >   e. <

10. a. .06, .6, .65
    b. 7.037, 7.307, 7.37
    c. 2.5, 2.505, 2.55

11. a. .9
    b. 11.43
    c. 3.142

PRACTICE EXERCISE 105

1. .6   2. .3   3. .1   4. .9   5. .7

6. $\frac{5}{10}$ 7. $\frac{2}{10}$ 8. $\frac{8}{10}$ 9. $\frac{1}{10}$ 10. $\frac{4}{10}$

PRACTICE EXERCISE 106

1. .37   2. .463   3. .99   4. .04

5. .036   6. .7834   7. .0834   8. .001

9. $\frac{53}{100}$ 10. $\frac{25}{100}$ 11. $\frac{8}{100}$ 12. $\frac{32}{1,000}$

13. $\frac{205}{1,000}$ 14. $\frac{1,034}{10,000}$ 15. $\frac{78}{10,000}$ 16. $\frac{5}{1,000}$

PRACTICE EXERCISE 107

1. nine-tenths   2. twenty-seven hundredths

3. five-hundredths   4. six and one-tenth

5. three and one hundred forty-one thousandths   6. fifty-one ten-thousandths

7. five and one-quarter hundredths   8. seven and two-hundredths

9. 6.09   10. .003   11. 3.5   12. 312.04

13. $.04\frac{3}{4}$ 14. .093   15. 7.0005   16. 40.43

17. .9   18. .25   19. .207   20. 3.014

21. 37.03   22. 900.009

PRACTICE EXERCISE 108

1. $\frac{5}{10} = \frac{1}{2}$ 2. $\frac{25}{100} = \frac{1}{4}$ 3. $\frac{625}{1,000} = \frac{5}{8}$

4. $\frac{75}{100} = \frac{3}{4}$ 5. $\frac{34}{100} = \frac{17}{50}$ 6. $\frac{125}{1,000} = \frac{1}{8}$

7. $\frac{9}{10}$ 8. $\frac{12}{100} = \frac{3}{25}$ 9. $\frac{8}{100} = \frac{2}{25}$

10. $3\frac{7}{10}$ 11. $5\frac{3}{4}$ 12. $12\frac{1}{10}$

13. $6\frac{3}{8}$ 14. $1\frac{7}{8}$ 15. $123\frac{1}{2}$

PRACTICE EXERCISE 109

1. .430
2. .040
3. .100
4. .56 and .560
5. .05, .05000, .050
6. .82
7. .95
8. .71
9. .04
10. .1
11. 4.01
12. .07 < .077 < .7
13. 3.001 < 3.01 < 3.1
14. 1.03 < 1.3 < 1.31
15. .0001 < .001 < .1

PRACTICE EXERCISE 110

1. .3
2. 2.4
3. 3.1
4. .7
5. 4.35
6. .77
7. 2.80
8. .87
9. 3.142
10. .678
11. 5.556

You have come a long way since you began to study and relearn the mathematics contained in this book. It has been fairly difficult for you to get to this point. You should be very proud of yourself. You still have a way to go before you can say you are proficient in mathematics. You have started on that climb; keep going!

11
adding and subtracting decimals
adding decimals

You have completed the introduction to decimals and have seen the relationship between the decimal form and the fraction form of a number. You have been successful in this area and you are ready to start the first fundamental operation associated with decimals, adding decimals.

Before you begin, though, take the pretest which follows, to see how much of the material covered in this topic is familiar to you.

See What You Know And Remember — Pretest 11A

Do as many problems as you can. Try as hard as you can. Some may be harder than others for you. Write each answer in the space provided.

| 1.   .2
     +.7 | 2.   .15
     +.04 | 3.   .76
     +.23 |

1.   .2
      +.7

2.   .15
      +.04

3.   .76
      +.23

4.   .07
      +.02

5.   .8
      +.6

6.   .3
      .5
      +.7

7.   3.6
      +2.1

8.   2.134
      + .35

9.   13.018
      + 7.703

10.   1.3 + 9.2 + 13 =

11.   4.5
      .07
      +20.457

12.   .013
      +.006

13. $.95 + $1.48 =

14. Bob went to the supermarket on Thursday to buy some bread and butter. The butter cost $.53 and the bread cost $.59. What was the total cost of his purchase?

15. How many miles did Sonia drive her car if she drove 14 mi. on Monday, 23.8 mi. on Tuesday, 5.3 mi. on Wednesday, and 3.25 mi. on Thursday?

16. The Capital Men's Shop had a sale on men's clothing. A suit sold for $74.95, a hat for $8.00, shoes for $16.99, and shirts for $13.00 each. How much money will it cost if you bought a suit, a pair of shoes, and one shirt?

Now turn to page 167 to check your answers. Add up all that you had correct.

A Score of	Means That You
15–16	Did very well. You can proceed to the second half of this chapter, "Subtracting Decimals."
13–14	Know this material except for a few points. Read the sections about the ones you missed.
11–12	Need to check carefully on the sections you missed.
0–10	Need to work with this part of the chapter to refresh your memory and improve your skills.

11.1 Adding Decimal Fractions

When adding decimal fractions, place the decimals in vertical columns, with each decimal point directly underneath the others. Place the decimal point in the answer directly below the decimal point in the addends and add the number as you would add whole numbers.

Example

What is the overall length of the steel template drawn below?

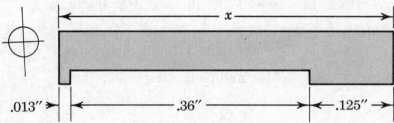

Solution

Problem	Step 1	Step 2
.013 + .360 + .125 =	.013 .360 + .125	.013 .360 + .125 .498 in.
	Place the decimals in vertical columns, lining up the decimal points directly below those of the addends.	Add the numbers.

Thus: .013 in. + .360 in. + .125 in. = .498 in.

Practice Exercise 111

Add these examples. Answers to this test and subsequent practice exercises begin on p. 167.

| 1. | .8
+.1 | 2. | .17
+.21 | 3. | .86
+.04 | 4. | .06
+.45 |

| 5. | .325
.242
+.051 | 6. | .08
.14
.23
+.46 | 7. | .504
+.308 | 8. | .46
+.30 |

What is the significance of zeros in decimal fractions?

Zeros to the right of a decimal fraction *do not* affect the value of the number. For instance:

$.5 or $.50 are both 50 cents.

Zeros to the right of the decimal point and to the left of the other digits in the decimal fraction *do* affect the value of the number. For instance:

$.01 or $.1 represent two different quantities.
$.01 = 1 penny while $.1 = 1 dime.

The zero is used as a place holder and cannot be omitted.

Example

The thickness of a piece of paper is .025 in. and the thickness of a strand of hair is .019 in. If one is placed on top of the other, how much will be their combined thickness?

Solution

Problem	Step 1	Step 2
.025 + .019 =	.025 +.019 .	.025 +.019 .044 in.
	Line up the fractions by the decimal points. Place the decimal point in the answer below that of the addends.	Add the numbers. Be sure to include the zero. Remember: .044 ≠ (is not equal to) .44.

Thus: .025 in. + .019 in. = .044 in.

Practice Exercise 112

Add. Be sure to include the zero if necessary.

1. .05
    +.03
    ─────

2. .037
    +.045
    ─────

3. .006
    +.045
    ─────

4. .0013
    .0121
    +.0504
    ──────

5. .009
    .027
    .031
    +.016
    ─────

6. .04
    .02
    +.05
    ─────

Sometimes, decimal fractions to be added contain different numbers of digits. The numbers will then be uneven on the right-hand side of the number when they are lined up in vertical columns. This is illustrated in the following example:

Example

Find the sum of .3205, .27, and .121.

Solution

Problem	Step 1	Step 2	Step 3
.3205 + .27 + .121 =	.3205 .27 +.121 ─────	.3205 .2700 +.1210 ────── .7115	.3205 .27 +.121 ───── .7115
	Line up the decimals and place the decimal point in the answer.	Add up the numbers. You can annex zeros to the right of each number without changing its value. Thus: .27 = .2700 .121 = .1210	OR Since the zeros are understood to be there, just add the numbers without annexing the zeros.

Thus: .3205 + .27 + .121 = .7115.

Practice Exercise 113

Add.

1.   .1034
     +.246

2.   .037
       .03
       .41
     +.002

3.   .274
       .29
     +.16

4.   .345
       .4
       .05
     +.19

5.   .48
       .1
     +.332

6.   .1
       .793
     +.07

11.2 Adding Mixed Decimals

Example

The Smith family plans a trip across country this summer and wishes to travel no more than 300 mi. each day. Checking the mileage between cities, they see that the mileage from Jackson to Lakeland is 151.2 miles, from Lakeland to Netcong is 87.3 miles, and from Netcong to Erie is 43.6 miles. Can the Smiths travel from Jackson to Erie and still be within their 300-mile limit?

Solution

Adding mixed decimals is the same as adding decimal fractions. Place the addends in vertical columns so that the decimal points are lined up one below the other. You can see that the whole numbers lie to the left of the decimal point and the decimal fractions to its right. Look at this example:

Problem	Step 1	Step 2
151.2 + 87.3 + 43.6 = ?	151.2 87.3 + 43.6 ————	151.2 87.3 + 43.6 ———— 282.1 mi.
	Line up the addends and place the decimal point in the answer.	Add the numbers.

Thus: 151.2 mi. + 87.3 mi. + 43.6 mi. = 282.1 mi.

This sum can be verified by changing each of the decimals to fractions and then adding.

Jackson to Lakeland $\qquad$ $151.2 = 151\dfrac{2}{10}$

Lakeland to Netcong $\qquad$ $87.3 = 87\dfrac{3}{10}$

Netcong to Erie $\qquad$ $43.6 = 43\dfrac{6}{10}$

$$282.1 \qquad 281\dfrac{11}{10} = 281 + 1\dfrac{1}{10} =$$

$$282\dfrac{1}{10} = 282.1 = 282.1 \text{ mi.}$$

And another —

Example

Find the sum of .73, .67, .2, and .259.

Solution

Problem	Step 1	Step 2
.73 + .67 + .2 + .259 =	.73 .67 .2 .259 ——— .	.73 .67 .2 .259 ——— 1.859
	Line up the addends and place the decimal point in the answer.	Add the numbers.

Thus: .73 + .67 + .2 + .259 = 1.859.

And another —

Example

Find the sum of 403.2, 45.96, 16, and .829.

Solution

Problem	Step 1	Step 2
403.2 + 45.96 + 16 + .829 =	403.2 45.96 16. + .829	403.2 45.96 16. + .829 465.989
	Line up the addends and place the decimal point in the answer. Remember that the number 16 is a whole number and the decimal point is at its far right.	Add the numbers.

Thus: 403.2 + 45.96 + 16 + .829 = 465.989.

Practice Exercise 114

Add.

| 1. | .8
+.4 | 2. | 3.56
+4.73 | 3. | 32.865
+ 2.97 | 4. | 52.985
+   .06 |

Arrange these problems in column form before adding. Be sure to place the decimal points one under the other.

5.   .345 + .4 + .05 + 1.9 =   6.   5.8 + 10 + .432 =

7.   \$72.87 + \$3.95 + \$1.99 =   8.   103.456 + 7.932 + 19.57 + .12 =

11.3   Word Problems Requiring The Addition Of Decimals

Adding decimals occurs frequently in various industries. One such problem is illustrated on the following page.

Example

Mr. Castro works in a metal shop. One day he received a blueprint drawing of a metal rod, pictured below, which he had to make.

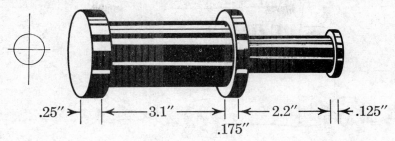

He had to find the overall length of the metal rod. What is the length?

Solution

To find the length of the metal rod he had to find the sum of its parts. Thus:

Problem	Step 1	Step 2
.25 + 3.1 + .175 + 2.2 + .125 =	.25 3.1 .175 2.2 + .125 .	.25 3.1 .175 2.2 + .125 5.850 or 5.85 in.
	Line up the addends and place the decimal point in the answer.	Add the numbers.

Thus: .25 in. + 3.1 in. + .175 in. + 2.2 in. + .125 in. = 5.85 in.

Practice Exercise 115

1. Mr. O'Leary is a traveling salesman and in one week covered these distances. On Monday he traveled 73.4 mi.; Tuesday, 55.8 mi.; Wednesday, 113.9 mi.; Thursday, 93.7 mi.; and Friday, 27.5 mi. Find the total number of miles he traveled during the week.

2. A driver was permitted to carry a maximum truck load of 1,500 lbs. He loaded 7 crates weighing 138.4 lbs., 258 lbs., 212.9 lbs., 371.6 lbs., 157.9 lbs., 197.2 lbs., and 164 lbs.

   a. Had he exceeded the maximum truck load of 1,500 lbs.?

b. If not, what was the weight of his load?

3. Ms. Chung is a social worker and receives a mileage allowance for her car. In one month she traveled the following weekly distances: 383.7 mi., 412.4 mi., 139.9 mi., and 273.8 mi. How many miles did she travel during the month?

4. Jenny shopped in a supermarket and bought the following priced items: $.89, $1.29, $.59, $.69, $.29, and $4.43. How much money does Jenny need to pay for her purchases?

5. A rectangular plot of land measures 43.87 ft. along its length and 31.9 ft. along its width. What is the perimeter of this plot of land?

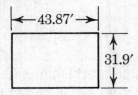

6. What is the length of this rod?

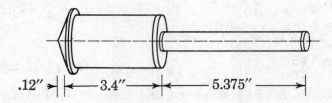

7. Find the **perimeter** of a triangle drawn below.

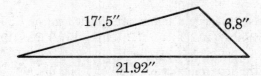

164

8. A sales slip was made for the following items. What was the total charge?

1 tennis shirt	$ 4.98
1 tennis short	12.98
1 can of balls	1.99
1 tennis racket	19.95

9. What is the total width of four books whose individual widths are 2.73 in., 0.67 in., 1.2 in., and 3.259 in.?

10. The world's indoor record for the two-mile run is 8 min. 13.2 sec. Mike ran the same distance in 18.9 sec. more. What was his total time?

Review Of Important Ideas

Some of the most important ideas in this part of Chapter 11 were:

 Adding decimals is similar to adding whole numbers.

 The decimal point separates the whole number portion of the number from the fractional part.

 To add decimal numbers —
1. Place the decimals in vertical columns with the decimal points directly underneath each other.
2. Add the numbers together as you would add whole numbers.
3. Place the decimal point in the answer directly below the decimal point in the addends.

Check What You Have Learned

The following test lets you see how well you learned the ideas thus far presented in Chapter 11.

Posttest 11A

Write each answer in the space provided.

1. .6
   +.3
   ‾‾‾

2. .12
   +.06
   ‾‾‾‾

3. .67
   +.32
   ‾‾‾‾

4.   .03
    +.06

5.   .9
    +.8

6.   .7
    .6
   +.4

7.   6.3
   +1.2

8.   2.431
   + .35

9.   14.017
   + 7.704

10. 1.4 + .86 + 14 =

11. 8.37
    .03
    4.9
   + .002

12. .0301
   +.0006

13. $2.59 + $.48 =

14. Martin went to the delicatessen on Tuesday to buy cole slaw and a carton of milk. The cole slaw cost $.59 and the container of milk cost $.87. What was the total cost of his purchase?

15. How many miles did Michael ride his bicycle if he rode 8.3 miles on Monday, 4.1 miles on Tuesday, 12.7 miles on Wednesday, and 2.46 miles on Thursday?

16. The Cape Cod Hardware Store had a second anniversary sale.

	Sale Price	Regular Price
Garden cart	$16.95	$18.88
Garden fencing	.89	2.29
Garden shovel	3.69	4.49
Cape Cod white picket fence	1.98	2.29
Garden fence kit	9.97	11.95
Plastic sprinkler can	1.95	2.69

Mrs. Maxson purchased a garden cart, a garden shovel, a plastic sprinkler can, and a garden fence kit. What was the total amount of her purchase?

```
ANSWERS AND EXPLANATIONS
TO POSTTEST 11A
```

1.   .6
   +.3
    .9

2.   .12
   +.06
    .18

3.   .67
   +.32
    .99

4.   .03
   +.06
    .09

5. .9 +.8 ――― 1.7	6. .7 .6 +.4 ――― 1.7	7. 6.3 +1.2 ――― 7.5	8. 2.431 + .35 ――― 2.781
9. 14.017 + 7.704 ――― 21.721	10. 1.4 .86 +14. ――― 16.26	11. 8.37 .03 4.9 + .002 ――― 13.302	12. .0301 +.0006 ――― .0307
13. $2.59 + .48 ――― $3.07	14. $ .59 + .87 ――― $1.46	15. 8.3 4.1 12.7 + 2.46 ――― 27.56 mi.	16. $16.95 3.69 1.95 + 9.97 ――― $32.56

A Score of	Means That You
15–16	Did very well. You can move on to the second part of the chapter, "Subtracting Decimals."
13–14	Know the material except for a few points. Reread the sections about the ones you missed.
11–12	Need to check carefully the sections you missed.
0–10	Need to review this part of the chapter again to refresh your memory and improve your skills.

Questions	Are Covered in Section
1–4, 12, 13	11.1
5–11	11.2
14–16	11.3

ANSWERS FOR CHAPTER 11 — A

PRETEST 11A

1. .9	2. .19	3. .99	4. .09
5. 1.4	6. 1.5	7. 5.7	8. 2.484
9. 20.721	10. 23.5	11. 25.027	12. .019
13. $2.43	14. $1.12	15. 46.35 mi.	16. $104.94

PRACTICE EXERCISE 111

| 1. .9 | 2. .38 | 3. .90 = .9 | 4. .51 |
| 5. .618 | 6. .91 | 7. .812 | 8. .76 |

PRACTICE EXERCISE 112

1. .08	2. .082	3. .051	4. .0638
5. .083	6. .11		

PRACTICE EXERCISE 113

1. .3494	2. .479	3. .724	4. .985
5. .912	6. .963		

PRACTICE EXERCISE 114

1. 1.2	2. 8.29	3. 35.835	4. 53.045
5. 2.695	6. 16.232	7. $78.81	8. 131.078

PRACTICE EXERCISE 115

1.
```
    73.4
    55.8
   113.9
    93.7
 +  27.5
 ───────
   364.3 mi.
```

2.
```
   138.4
   258.
   212.9
   371.6
   157.9
   197.2
 + 164.
 ───────
 1,500.0 lbs.
```

a. No
b. 1,500 lbs.

3.
```
   383.7
   412.4
   139.9
 + 273.8
 ───────
 1,209.8 mi.
```

4.
```
 $  .89
    1.29
     .59
     .69
     .29
 +  4.43
 ───────
  $8.18
```

5. $P = 43.87 + 43.87 + 31.9 + 31.9$
   $P = 151.54$ ft.

6.
```
    .12
   3.4
 + 5.375
 ───────
   8.895 in.
```

7. $P = 17.5 + 6.8 + 21.92$
   $P = 46.22$ in.

8. $39.90

9.
```
   2.73
   0.67
   1.2
 + 3.259
 ──────
   7.859 in.
```

10.
```
 8 min. 13.2 sec.
         18.9 sec.
 ────────────────
 8 min. 32.1 sec.
```

168

Subtracting decimals

To answer the question "How much remains after part is used?" you need to subtract. Change from a $10 bill after a purchase is made depends on subtraction. Finding the difference between any two quantities is another example of the operation. It is extremely important to know how to subtract decimals.

The pretest that follows will allow you an opportunity to see how good your skills are in subtracting decimals.

See What You Know And Remember — Pretest 11B

Some of these problems are more difficult than others. Do as many as you can. Write each answer in the space provided.

1. $\begin{array}{r} .8 \\ -.2 \\ \hline \end{array}$

2. $\begin{array}{r} .69 \\ -.41 \\ \hline \end{array}$

3. $7.6 - 3 =$

4. $\begin{array}{r} .39 \\ -.32 \\ \hline \end{array}$

5. $\begin{array}{r} 9.78 \\ -7.65 \\ \hline \end{array}$

6. $\begin{array}{r} 4.09 \\ -1.03 \\ \hline \end{array}$

7. $\begin{array}{r} 8.5 \\ -2.6 \\ \hline \end{array}$

8. $58.9 - 4.6 =$

9. $5 - .7 =$

10. $\begin{array}{r} 6.84 \\ -\ .58 \\ \hline \end{array}$

11. $\begin{array}{r} 91.3 \\ -73.9 \\ \hline \end{array}$

12. $\begin{array}{r} .503 \\ -.39 \\ \hline \end{array}$

13. $\begin{array}{r} 6.51 \\ -2.47 \\ \hline \end{array}$

14. $\begin{array}{r} .084 \\ -.053 \\ \hline \end{array}$

15. $\begin{array}{r} 1.255 \\ -\ .165 \\ \hline \end{array}$

16. $1.03 - .75 =$

17. $\begin{array}{r} 39.29 \\ -\ 4.7 \\ \hline \end{array}$

18. $5 - 3.2 =$

19. $19.8 - .32 =$

20. $.73 - .648 =$

21. Ramona had a piece of wood 8.75 ft. long. If she cut off a piece 3.25 ft. long to use for a shelf, how long was the piece which remained?

22. In the 1896 Olympics, Thomas E. Burke won the 100-meter dash in 12.0 sec. In the 1968 Olympics, James Hines won the same race in 9.9 sec. How much faster than Burke was Hines?

23. Edna weighed 117.65 lb. when summer began. When summer ended in September, she weighed 121.5 lb. How many pounds did Edna gain?

24. The gas tank in Mrs. Watts' car holds 20 gal. It took 16.3 gal. to fill the tank. How much gas was in the tank before it was filled?

25. Carol bought a book for $3.95 and a game for $2.35. She gave the clerk a $20 bill. How much change did she get?

Now turn to the end of the chapter to check your answers. Add up all that you had correct.

A Score of	Means That You
23–25	Did very well. You can proceed to Chapter 12.
20–22	Know the material except for a few points. Read the sections about the ones you missed.
17–19	Need to check carefully on the sections you missed.
0–16	Need to work with the chapter to refresh your memory and improve your skills.

Questions	Are Covered in Section
1, 2, 4, 12, 14, 20	11.4
3, 5–11, 13, 15–19	11.5
21–25	11.6

11.4 Subtracting Decimal Fractions

Example

The thickness of a piece of paper is approximately .025 in. How much larger is this than the thickness of a strand of hair measuring .019 in.?

Solution

You subtract .019 in. from .025 in. to get the solution to this example. To subtract (.025 − .019), you place the decimals in vertical columns with one decimal point directly underneath the other. Place the decimal point in the answer, or *difference*, directly below the decimal points in the *minuend* and *subtrahend*. Subtract the numbers as you would subtract whole numbers. See the following illustration:

Problem	Step 1	Step 2
.025 − .019 =	.025 −.019 _____ .	1_1 .025 −.019 _____ .006 in.
	Place the decimals in vertical columns, lining up the decimal points. Place the decimal point in the answer directly below the other decimal points.	Subtract the numbers. Be sure to include the zeros. (Remember: .006 ≠ [is not equal to] .6.)

Thus: .025 in. − .019 in. = .006 in.

Practice Exercise 116

Subtract. Be sure to indicate zeros when necessary.

1.   .9
    −.3

2.   .48
    −.27

3.   .311
    −.287

4.   .431
    −.278

5.   .1134
    −.0879

6.   .373
    −.298

7.   .302
    −.288

8.   .151
    −.146

9.   .995
    −.495

Example

Caridad won the race in .53 min. Her closest competitor ran the same race in .549 min. How much faster was Caridad?

Solution

Problem	Step 1	Step 2	Step 3
.549 − .53 =	.549 −.53 .	.549 −.530 .	.549 −.530 ——— .019 min.
	Line up the decimal points so that one is directly below the other. Place the decimal point in the answer below the others.	.53 = .530 (since zeros to the right of the decimal fraction do not change its value)	Subtract the numbers. Be sure to include the zero.

Thus: .549 min. − .53 min. h-1 .019 min.

Example

Find the difference between .54 and .259.

Solution

Problem	Step 1	Step 2	Step 3
.54 − .259 =	.54 −.259 .	.540 −.259 .	.540 −.259 ——— .281
	Line up the decimal points so that one is directly below the other. Place the decimal point in the answer below the others.	.54 = .540 (since zeros to the right of the decimal fraction do not change its value)	Subtract the numbers.

Thus: .54 − .259 = .281.

172

Practice Exercise 117

Subtract.

1.   .547
    −.29
    ―――

2.   .43
    −.278
    ―――

3.   .647
    −.19
    ―――

4.   .6
    −.273
    ―――

5.   .187
    −.099
    ―――

6.   .07
    −.0691
    ―――

7.   .482
    −.27
    ―――

8.   .5644
    −.38
    ―――

9.   .302
    −.28
    ―――

11.5   Subtracting Mixed Decimals

Mixed decimals are subtracted in the same way as decimal fractions.

Example

Mr. Gonzalez had a credit of $8.95 in a men's clothing store. He bought a shirt for $4.99, a tie for $2.99, and socks for $2.98. How much additional money must he pay?

Solution

Mr. Gonzalez received a sales slip with the following computations:

A & R MEN'S STORE	
1 shirt	$ 4.99
1 tie	2.99
2 pair of socks	2.98
	$10.96
Credit	8.95
To Be Paid	$ 2.01

From this information on the sales slip, Mr. Gonzalez determined that he had to pay an additional amount of $2.01 to cover the cost of his purchases. See the following subtraction example:

Problem	Step 1	Step 2
$10.96 − $8.95 =	$10.96 − 8.95 _____ .	$10.96 − 8.95 _____ $ 2.01
	Line up the decimal points and place the decimal point in the answer.	Subtract the numbers. Be sure to include the zero in the answer.

Thus: $10.96 − $8.95 = $2.01.

Subtraction examples are checked by adding the *difference* to the *subtrahend*. This sum must correspond to the number in the *minuend*. In this example you have:

$10.96   (minuend)   and $ 8.95   (subtrahend)
− 8.95   (subtrahend)   + 2.01   (difference)
_____   _____
$ 2.01   (difference) $10.96   (minuend)

Thus, you see that the $2.01 answer checks.

Practice Exercise 118

Subtract.

1. $5.43
   − 2.59
   _____

2.   3.11
   − 2.87
   _____

3.   3.73
   − 2.98
   _____

4.   56.44
   − .38
   _____

5.   38.16
   − 9.47
   _____

6.   3.002
   − 2.97
   _____

7.   2.83
   − .5737
   _____

8. $113.43
   −   87.95
   _____

9.   1.87
   − .99
   _____

Example

Beulah left work with $5 and spent $3.79 on her way home. How much money did she have when she arrived home?

Solution

Problem	Step 1	Step 2	Step 3
$5 − $3.79 =	$5. − 3.79 ⎯⎯⎯ .	$5.00 − 3.79 ⎯⎯⎯ .	$5.00 − 3.79 ⎯⎯⎯ $1.21
	Line up the decimals and place the decimal point in the answer. The decimal point in 5 is on its right: 5.	Annex 2 zeros to the right of the decimal point: 5. = 5.00	Subtract the numbers.

Thus: $5 − $3.79 = $1.21.

And another —

Example

Find the remainder when .6 is subtracted from 4.

Solution

Problem	Step 1	Step 2	Step 3
4 − .6 =	4. − .6 ⎯⎯⎯ .	4.0 − .6 ⎯⎯⎯ .	4.0 − .6 ⎯⎯⎯ 3.4
	The decimal point in a whole number is on its right: 4.	Annexing zeros to the right of the decimal point doesn't change the value of the number: 4. = 4.0	Subtract the numbers.

Thus: 4 − .6 = 3.4.

Practice Exercise 119

Subtract. Place the decimal point in the numbers where it is needed. Remember to include zeros when necessary.

1. $4 – $1.23 =

2. 6 – .8 =

3. 4.9 – 2 =

4. 5.87 – .9 =

5. 113.4 – 87 =

6. 37 – 4.21 =

7. $6 – $1.95 =

8. Subtract .273 from 6.

9. From $10 subtract $6.91.

11.6   Word Problems Requiring The Subtraction Of Decimals

The subtraction of decimals is necessary at many different times in our everyday life. One such example is illustrated below.

Example

The radio announcer on Station WGLX reported that the barometer had fallen from 30.2 in. to 28.8 in. since 4:00 p.m. How many inches had it dropped?

Solution

This example requires you to find the difference in the two barometer readings. The operation is subtraction, so subtract 28.8 in. from 30.2 in. See the following example.

176

Problem	Step 1	Step 2
30.2 − 28.8 =	$\begin{array}{r} 30.2 \\ -28.8 \\ \hline . \end{array}$	$\begin{array}{r} 29\ 1 \\ \cancel{30.2} \\ -28.8 \\ \hline 1.4 \text{ in.} \end{array}$
	Line up the decimals and place the decimal point in the answer.	Subtract the decimals.

Thus: 30.2 in. − 28.8 in. = 1.4 in.

Practice Exercise 120

1. Two sheets of aluminum have thicknesses of .028 in. and .032 in. By how much do they differ in size?

2. Before the Reeds left New York for Cape Cod, the automobile odometer read 4,543.7 mi. Upon their arrival, the odometer read 4,830.4 mi. How many miles is it from New York to Cape Cod?

3. Ms. Ramirez purchased a jacket for $9.95, a hat for $4.95, and a pair of shoes for $13.95. How much money did she receive in change if she gave the cashier $30?

4. The winner in a horse race ran the mile in 1.46 min. The second-place horse completed the mile in 1.51 min. By how many minutes faster was the winner?

5. Mr. Garcia measured a metal rod and found its length to be 5.85 in. If metal shrinks .005 in. during the winter months, what will its new length be?

6. A woman's club in Baltimore collected $564.75 from a garage sale. What was the profit if their expenses were $35.95?

7. If the perimeter of a triangular plot of ground is 34.671 ft. and two of its sides measure 12.7 ft. and 15.386 ft., find the length of its third side.

177

8. Farmer Appleby owns two different-shaped fields. One field is triangular and its dimensions are 138.2 yd., 170.6 yd., and 92.4 yd. The other is rectangular in shape and its dimensions are 89.6 yd. by 47.8 yd.

   a. What is the perimeter of the triangular field?

   b. What is the perimeter of the rectangular field?

   c. Which field has the longer perimeter?

   d. How many yards larger is it?

9. The gasoline tank in Mr. Alfaro's automobile holds 24 gal. If Mr. Alfaro fills the tank with 13.4 gal., how many gallons were in the tank before he filled it?

10. In the diagram below, find the missing dimension x.

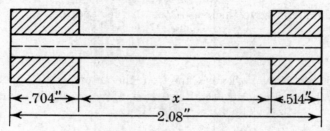

Review Of Important Ideas

Some of the most important ideas in this part of Chapter 11 were:

 Subtracting decimals is almost identical to subtracting whole numbers.

 The decimal point is always placed to the right of a whole number.

 Annexing zeros to the right of the decimal fraction will not change its value.

 To subtract decimals —
1. Place the subtrahend under the minuend so that the decimal points are directly beneath one another.
2. Place the decimal point in the difference directly below the decimal points in the minuend and subtrahend above.
3. Subtract the numbers as you would whole numbers.

Check What You Have Learned

Good luck on this exercise. These examples will test your understanding of subtracting decimals.

Posttest 11B

Write each answer in the space provided.

1.
```
   .9
 − .3
 ────
```

2.
```
   .78
 − .36
 ─────
```

3. $8.4 - 5 =$

4.
```
   .59
 − .55
 ─────
```

5.
```
   8.79
 − 5.67
 ──────
```

6.
```
   9.04
 − 1.03
 ──────
```

7.
```
   5.6
 − 2.8
 ─────
```

8. $85.9 - 4.6 =$

9. $7 - .5 =$

10.
```
   6.88
 −  .59
 ──────
```

11.
```
   31.1
 − 13.9
 ──────
```

12.
```
   .407
 − .29
 ─────
```

13.
```
   9.53
 − 6.49
 ──────
```

14.
```
   .048
 − .037
 ──────
```

15.
```
   1.345
 −  .265
 ───────
```

16. $3.04 - .79 =$

17.
```
   28.37
 −  5.6
 ───────
```

18. $7 - 3.8 =$

19. $27.5 - .38 =$

20. $.83 - .752 =$

21. Bob had an oil truck which held 350 gal. of oil. It took 273.2 gal. to fill the tank. How much oil was in the tank before it was filled?

22. Rafael cut a piece of wood 5.25 ft. long from a longer piece measuring 15.75 ft. How long a piece of wood remained?

23. Maria bought a record for $4.99 and a book for $2.75. She gave the clerk a $10 bill. How much change did she get?

24. Last year's winner ran the women's 60-yard dash in 7.0 sec. If the Madison Square Garden record is 6.6 sec., how many seconds slower was last year's winner?

25. At the close of the year Daisy weighed 98.5 lb. and in February she weighed 103.2 lb. How many pounds did she gain during January?

```
┌─────────────────────────────┐
│   ANSWERS AND EXPLANATIONS   │
│      TO POSTTEST 11B         │
└─────────────────────────────┘
```

1. .9
   −.3
   ―――
    .6

2. .78
   −.36
   ―――
    .42

3. 8.4
   −5.0
   ―――
    3.4

4. .59
   −.55
   ―――
    .04

5. 8.79
   −5.67
   ―――
    3.12

6. 9.04
   −1.03
   ―――
    8.01

7. 5.6
   −2.8
   ―――
    2.8

8. 85.9
   − 4.6
   ―――
    81.3

9. 7.0
   − .5
   ―――
    6.5

10. 6.88
   − .59
   ―――
    6.29

11. 31.1
   −13.9
   ―――
    17.2

12. .407
   −.29
   ―――
    .117

13. 9.53
   −6.49
   ―――
    3.04

14. .048
   −.037
   ―――
    .011

15. 1.345
   − .265
   ―――
    1.08

16. 3.04
   − .79
   ―――
    2.25

17. 28.37
   − 5.6
   ―――
    22.77

18. 7.0
   −3.8
   ―――
    3.2

19. 27.50
   − .38
   ―――
    27.12

20. .830
   −.752
   ―――
    .078

21. 350.0
   −273.2
   ―――
    76.8 gal.

22. 15.75
   − 5.25
   ―――
    10.5 ft.

23. $4.99
   + 2.75
   ―――
    $7.74

    $10.00
   − 7.74
   ―――
    $ 2.26

24. 7.0
   −6.6
   ―――
    .4 sec.

25. 103.2
   − 98.5
   ―――
    4.7 lb.

A Score of	Means That You
23–25	Did very well. You can move on to Chapter 12.
20–22	Know this material except for a few points. Reread the sections about the ones you missed.
17–19	Need to check carefully the sections you missed.
0–16	Need to review this part of the chapter again to refresh your memory and improve your skills.

Questions	Are Covered in Section
1, 2, 4, 12, 14, 20	11.4
3, 5–11, 13, 15–19	11.5
21–25	11.6

ANSWERS FOR CHAPTER 11 — B

PRETEST 11B

1.	.6	2.	.28	3.	4.6	4.	.07	5.	2.13
6.	3.06	7.	5.9	8.	54.3	9.	4.3	10.	6.26
11.	17.4	12.	.113	13.	4.04	14.	.031	15.	1.090
16.	.28	17.	34.59	18.	1.8	19.	19.48	20.	.082
21.	5.5 ft.	22.	2.1 sec.	23.	3.85 lb.	24.	3.7 gal.	25.	$13.70

PRACTICE EXERCISE 116

1.	.6	2.	.21	3.	.024
4.	.153	5.	.0255	6.	.075
7.	.014	8.	.005	9.	.5

PRACTICE EXERCISE 117

1.	.257	2.	.152	3.	.457
4.	.327	5.	.088	6.	.0009
7.	.212	8.	.1844	9.	.022

PRACTICE EXERCISE 118

1.	$2.84	2.	.24	3.	.75
4.	56.06	5.	28.69	6.	.032
7.	2.2563	8.	$25.48	9.	.88

PRACTICE EXERCISE 119

1. $2.77
4. 4.97
7. $4.05

2. 5.2
5. 26.4
8. 5.727

3. 2.9
6. 32.79
9. $3.09

PRACTICE EXERCISE 120

1.
```
  .032
 −.028
 ─────
  .004 in.
```

2.
```
  4,830.4
 −4,543.7
 ────────
   286.7 mi.
```

3.
```
 $ 9.95
   4.95
 + 13.95
 ───────
 $28.85
```
```
 $30.00
 − 28.85
 ───────
 $ 1.15
```

4.
```
  1.51
 −1.46
 ─────
  .05 min.
```

5.
```
  5.850
 − .005
 ──────
  5.845 in.
```

6.
```
 $564.75
 − 35.95
 ───────
 $528.80
```

7.
```
  12.7
 +15.386
 ──────
  28.086
```
```
  34.671
 −28.086
 ───────
   6.585 ft.
```

8. a. $P = 138.2 + 170.6 + 92.4$
   $P = 401.2$ yds.
   b. $P = 89.6 + 89.6 + 47.8 + 47.8$
   $P = 274.8$ yds.
   c. Triangular field
   d.
```
   401.2
 −274.8
 ──────
  126.4 yds.
```

9.
```
  24.0
 −13.4
 ─────
  10.6 gal.
```

10.
```
  .704
 + .514
 ──────
  1.218
```
```
  2.080
 −1.218
 ──────
   .862 in. = x
```

Multiplying decimals is another basic operation you must know if you are to be proficient with mathematics. You use multiplication to find the total cost when you purchase 2 quarts of milk at $.48 a quart or 3 cans of juice at $.43 a can. Cashiers are constantly required to use this skill.

To see how well you know this material, take the pretest which follows.

See What You Know And Remember — Pretest 12

Be careful doing these problems. Try to get as many correct as you can. Write each answer in the space provided.

1. .2
    × 4

2. 2.3
    × 3

3. 2.63
    × 2

4. .8
    ×.5

5. 4.03
    × 2

6. .197
    × 2

7. .23
    ×.4

8. 27
    ×.08

9. 4.6
    ×1.8
    _____

10. .67
    ×8.4
    _____

11. .78
    ×.96
    _____

12. 9.3
    ×.09
    _____

13. .704
    × .06
    _____

14. .29
    ×.07
    _____

15. a. .4 × 100 =

    b. 1.2 × 10 =

    c. 3.84 × 1,000 =

16. 32.6
    × 1.3
    _____

17. 48.7
    × .68
    _____

18. 8.02
    × .49
    _____

19. $59.13
    ×   78
    _____

20. Mr. Michaels bought a chicken which weighed 3.75 lb. and cost $.49 a pound. How much did he pay for the chicken?

21. Griselda decides to bake a chocolate layer cake. The recipe calls for .5 lb. of chocolate. If she triples the recipe, how many pounds of chocolate will she need?

22. A piece of paper is .008 in. thick. If a package contains 250 of these sheets, how thick is the package?

23. An automobile can travel 24 mi. on a gallon of gasoline. How far can it travel on .8 of a gallon?

Now turn to the end of the chapter to check your answers. Add up all that you had correct. Count by the number of separate answers, not by the number of questions. In this pretest there were 23 questions, but 25 separate answers.

A Score of	Means That You
23–25	Did very well. You can proceed to Chapter 13.
20–22	Know this material except for a few points. Read the sections about the ones you missed.
16–19	Need to check carefully on the sections you missed.
0–15	Need to work with this chapter to refresh your memory and improve your skills.

Questions	Are Covered in Section
1–3, 5, 6 8, 19	12.1
4, 7, 9–14, 16–18	12.2
15	12.3
20–23	12.4

12.1 Multiplying A Decimal By A Whole Number

Example

How much money does it cost to purchase 2 qt. of milk at $.48 a quart?

Solution

To answer that question requires the operation of either addition or multiplication. You could add the cost of each quart of milk or multiply the price of one quart by 2. These methods are shown below:

Addition	*Multiplication*

$$\begin{array}{r} \$.48 \\ +\ .48 \\ \hline \$.96 \end{array} \qquad \$.48 \text{ or } \frac{48}{100} \times 2 = \frac{96}{100} = .96$$

$$\begin{array}{r} \$.48 \\ \times\ 2 \\ \hline \$.96 \end{array}$$

And another —

Example

Find the product of .176 and 3.

Solution

Addition	Multiplication
.176	.176 or $\dfrac{176}{1,000} \times 3 = \dfrac{528}{1,000} = .528$
.176	
+.176	$\times\ 3$
———	———
.528	.528

To multiply a decimal by a whole number you multiply the numbers as you would multiply whole numbers. The decimal in the *product* contains *the same number of places as the number of decimal places in the multiplicand*. See the following example:

Example

How long must a metal rod measure if you wish to cut 6 rods of 5.85 in. each?

Solution

Problem	Step 1	Step 2
5.85 $\times\quad 6$	5.85 $\times\quad 6$ ——— 3510	5.85 (multiplicand) $\times\quad 6$ (multiplier) ——— 35.10 (product)
	Multiply the numbers: $585 \times 6 = 3510$	5.85 has 2 decimal places. Place the decimal point in 3510 counting 2 places from right to left. Thus: $35.10 = 35.1$

Thus: 5.85 in. $\times$ 6 = 35.1 in.

Practice Exercise 121

The products in problems 1–10 have been completed but the decimal point has been left out of the answer. Place the decimal point where it belongs.

1. $200 \times .05 = 1000$

2. $40 \times .30 = 1200$

3. $4 \times .5 = 20$

4. $4.62 \times 4 = 1848$

5. $.525 \times 6 = 3150$

6. $2 \times .2 = 4$

7. .50 × 5 = 250   8. 1.9 × 5 = 95

9. .2415 × 3 = 7245   10. .06 × 6 = 36

Find the product of each of the following examples. Place the decimal point in the answer.

11. .125 × 500 =   12. 5.25 × 4 =

13. 14.5 × 12 =   14. .4 × 5 =

15. 48.32 × 6 =   16. 5 × .96 =

17. 3 × .04 =   18. $14.38 × 7 =

19. $146.30 × 22 =   20. $62.50 × 8 =

12.2 Multiplying Decimal Numbers

Example

If a metal rod costs $.23 an inch, how many dollars would a rod 35.1 in. long cost?

Solution

To answer this question you multiply 35.1 in. by $.23, or

$$
\begin{array}{r}
35.1 \\
\times \$.23 \\
\hline
\end{array}
$$

Follow the same procedure as illustrated in the previous section. You multiply as you would with whole numbers. You must then place the decimal point in the product. To decide where to place the decimal point, let's do this same example by writing each decimal in fraction form and then multiplying. Thus:

$$35.1 \times \$.23 =$$

$$35\frac{1}{10} \times \frac{23}{100} =$$

$$\frac{351}{10} \times \frac{23}{100} = \frac{8073}{1000} = 8\frac{73}{1000}$$

$8.073 or $8.07 rounded off to the nearest cent

Since the product of 35.1 × .23 using fractions is 8.073, then this same product should be obtained if you use decimals. Thus:

$$
\begin{array}{rl}
35.1 & \text{(multiplicand)} \\
\times \ \ .23 & \text{(multiplier)} \\
\hline
8.073 & \text{(product)}
\end{array}
$$

The multiplicand is a *one*-place decimal, the multiplier is a *two*-place decimal, and the product is a *three*-place decimal. This leads to the conclusion that you place the decimal point in the product, counting from right to left, the same number of places as the total decimal places in the multiplicand and multiplier. See this illustration:

Problem	Step 1	Step 2
35.1 × .23	35.1 × .23 ――― 1053 702 ――― 8073	35.1 (one-place decimal) × .23 (two-place decimal) ――― 1053 702 ――― 8.073 (three-place decimal)
	Multiply the numbers: 351 × 23 = 8073	Place the decimal point in the product, counting 3 places from the right. Place it between the 8 and 0.

Thus: 35.1 × .23 = 8.073.

Notice that the decimal points do not need to be underneath one another before you can multiply. This is different from adding or subtracting. Check this method out on these two examples.

Example

Find the product of 4.56 × .12.

Solution (using fractions)

$$
4.56 \times .12 =
$$
$$
4\frac{56}{100} \times \frac{12}{100} =
$$
$$
\frac{456}{100} \times \frac{12}{100} = \frac{5472}{10000} = .5472
$$

or solution using decimals

Problem	Step 1	Step 2
4.56 × .12	4.56 × .12 912 456 5472	4.56 (two-place decimal) × .12 (two-place decimal) 912 456 .5472 (four-place decimal)
	Multiply the numbers: 456 × 12 = 5472	The sum of the decimal places in the multiplicand and multiplier is 4. Place the decimal point, starting from the right, 4 places to the left.

Thus: 4.56 × .12 = .5472.

And another —

Example

What is the product of .003 and .01?

Solution

Problem	Step 1	Step 2
.003 × .01	.003 × .01 3	.003 (three-place decimal) × .01 (two-place decimal) .00003 (five-place decimal)
	Multiply the numbers: 3 × 1 = 3	The sum of the decimal places in the multiplicand and multiplier is 5. Place 4 zeros to the left of 3 as placeholders. The result is .00003.

Thus: .003 × .01 = .00003.

Practice Exercise 122

The products in problems 1–10 have been completed but the decimal points have been left out of the answer. Place the decimal point where it belongs. Include zeros as place holders where necessary.

1. $4.62 \times .4 = 1848$

2. $5.25 \times .6 = 3150$

3. $.2415 \times .3 = 7245$

4. $.06 \times .6 = 36$

5. $48.32 \times .06 = 28992$

6. $.5 \times .96 = 480$

7. $.03 \times .04 = 12$

In examples 8–10 round off the answer to the nearest cent after placing the decimal point.

8. $\$14.38 \times .07 = \$10066$

9. $\$1,463 \times .222 = \$324786$

10. $\$62.50 \times 1.1 = \$68750$

Find the products in each of the following examples. Place the decimal point in the answer and add zeros when necessary.

11. $1.73 \times .8 = 1384$

12. $37.42 \times 3.6 = 134712$

13. $13.2 \times .43 = 5676$

14. $3.14 \times .16 = 5024$

15. $.037 \times .01 = 37$

In examples 16–20 round off the answer to the nearest cent after placing the decimal point.

16. $\$36.42 \times .08 = \$29136$

17. $\$149.78 \times .15 = \$224670$

18. $\$3,200 \times .055 = \$176000$

19. $\$225.72 \times 3.6 = \$812592$

20. $\$83.75 \times .042 = \$351750$

12.3 Special Multipliers Of 10, 100, 1000, Etc.

It is often necessary to multiply by the special numbers 10, 100, 1,000, etc. Short cuts save considerable amounts of time so look at these special multipliers in the following examples:

$$\begin{array}{r} 1.234 \\ \times\quad 10 \\ \hline 12.34\cancel{0} \end{array} \qquad \begin{array}{r} 1.234 \\ \times\quad 100 \\ \hline 123.4\cancel{0}\cancel{0} \end{array} \qquad \begin{array}{r} 1.234 \\ \times\quad 1,000 \\ \hline 1,234.\cancel{0}\cancel{0}\cancel{0} \end{array}$$

Notice that the decimal point in the multiplicand moved *one* place to the right when multiplying by *10,* *two* places to the right when multiplying by *100,* and *three* places to the right when multiplying by *1,000.* Therefore, when multiplying by the special numbers 10, 100, 1,000, etc., the decimal point in the multiplicand moves a number of places to the right equal to the number of zeros there are in the multiplier.

Thus, multiplying by 10, which contains one zero, moves the decimal point in the multiplicand one place to the right. Multiplying by 100, which contains two zeros, moves the decimal point in the multiplicand two places to the right. Look at the following examples which illustrate the short cut for the special multipliers of 10, 100, 1,000, etc.

Problem	Step 1
$14.32 \times 10 =$	$14.32 \times 10 = 14.3.2$ $= 143.2$
	Move the decimal point in the multiplicand 14.32 one place to the right since the multiplier 10 contains one zero.

Problem	Step 1
$.138 \times 10,000 =$	$.138 \times 10,000 =$ $.1380. = 1,380$
	Move the decimal in the multiplicand .138 four places to the right since the multiplier 10,000 contains four zeros. Annex a zero as a place holder.

Thus: $14.32 \times 10 = 143.2.$

Thus: $.138 \times 10,000 = 1,380.$

Practice Exercise 123

Multiply. The decimal point has been left out of the answer. Place the decimal point where it belongs. Put in zeros when necessary.

1. $10 \times 4.2 = 42$

2. $13.625 \times 100 = 13625$

3. $1.4 \times 10 = 14$

4. $1000 \times 2.4 = 24$

5. $.247 \times 10 = 247$

6. $3.1416 \times 100 = 31416$

7. $.62 \times 10000 = 62$

8. $10 \times .4 = 4$

9. $100 \times \$125 = \$125$

10. $\$3.47 \times 10 = \$347$

12.4 Word Problems Requiring The Multiplication Of Decimals

Example

Mr. Thomas fills his car with 14.3 gal. of gasoline costing 52.3¢ a gallon. How much does the gasoline cost him?

Solution

Each gallon of gasoline cost 52.3¢, so multiply to find the cost of 14.3 gal.

Problem	Step 1	Step 2	
52.3 ×14.3	52.3 × 14.3 ——— 1569 2092 523 ——— 74789	52.3 × 14.3 ——— 1569 2092 523 ——— 747.89¢	(one-place decimal) (one-place decimal) (two-place decimal)
	Multiply the numbers: 523 × 143 = 74789	Mark off the decimal point in the product, starting from the right, two places to the left.	

Thus: 52.3 × 14.3 = 747.89¢

Since 100 cents = 1 dollar, 747.89¢ = $7.4789, which rounded off to the nearest cent is $7.48.

Practice Exercise 124

1. Jane buys 3.5 yd. of material at the cost of $2.39 per yd. How much does the material cost?

2. John Lewis belongs to a pension plan in his union. Each week $2.97 is deducted from his paycheck to pay for this pension plan. How much does he contribute to the fund after a full year of work?   (Remember: 52 weeks = 1 year.)

3. Rain fell in the city this year at an average of .641 in. per month. How many inches of rain fell during the year?

4. Ms. Pearl knows that she can drive 22.6 mi. for each gallon of gasoline. If the capacity of her gasoline tank is 11.8 gal., then how far can she travel without stopping to refuel?

5. If it costs a hospital $1.28 per day to feed lunch to each of its patients, then what is the cost for 1,000 patients?

6. Find .07 of $48.45 rounded off to the nearest cent.

7. If a store owner sells 12 boxes of candy at 2 boxes for $.75, how much money did he receive for his candy?

8. Find the area of a rectangular plot of ground whose dimensions are 26.8 ft. by 31.37 ft.

9. The length of a circle is called the *circumference*. It can be found by multiplying the diameter, indicated on the diagram as 14 in., by 3.14, which is the value called *pi* (π). Find the circumference of this circle.

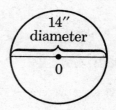

10. At $1.49 a pound, what is the cost of a chuck steak weighing 3.2 pounds?

Review Of Important Ideas

Some of the most important ideas in Chapter 12 were:

 Multiplying decimals is similar to multiplying whole numbers.

 To multiply a decimal by either a whole number or a decimal —
1. Multiply the numbers in the same way as you would whole numbers.
2. Find the sum of the decimal places in the multiplier and the multiplicand.
3. Place the decimal point in the product, counting from right to left, the same number of places as the sum of the decimal places in the multiplier and the multiplicand.

 To multiply by special numbers like 10, 100, 1,000, etc. —
1. Keep the same digits in the product as there are in the multiplicand.
2. Place the decimal point in the product as many places to the right of its position in the multiplicand as there are zeros in the multiplier. Add zeros as place holders when necessary.

Check What You Have Learned

This test will reveal how well you have learned the material in this chapter. Remember, it's accuracy, not speed, which is most important.

Posttest 12

Try hard for a satisfactory grade.

1. .3
    × 3

2. 3.2
    × 3

3. 3.53
    × 2

4. .7
    ×.6

5. 3.04
    × 2

6. .186
    × 3

7. .32
    × .4

8. 34
    ×.07

9. 6.4
    ×1.8

10. .76
    ×8.4

11. .87
    ×.69

12. 9.5
    ×.07

13. .407
    × .06

14. .92
    ×.07

15. a. .6 × 100 =

    b. 9.2 × 10 =

    c. 3.75 × 1,000 =

16. 44.8
    × 1.2

17. 84.7
    × .68

18. 9.02
    × .38

19. $39.43
    ×   87

20. Mr. Bates bought a steak weighing 2.7 lb. which cost $1.29 per lb. How much did he pay for the steak?

21. Cathy decides to cook a cheese fondue. The recipe calls for .5 cups of wine. If she triples the recipe, how much wine will she need?

194

22. A sheet of cardboard is .015 in. thick. If 150 sheets of cardboard are stacked one on top of the other, how thick is the stack?

23. An automobile can travel 18 mi. on a gallon of gasoline. How far can it travel on .6 of a gallon?

**ANSWERS AND EXPLANATIONS
TO POSTTEST 12**

1.   .3
     × 3
     .9

2.   3.2
     × 3
     9.6

3.   3.53
     × 2
     7.06

4.   .7
     ×.6
     .42

5.   3.04
     × 2
     6.08

6.   .186
     × 3
     .558

7.   .32
     × .4
     .128

8.   34
     ×.07
     2.38

9.   6.4
     × 1.8
     512
     64
     11.52

10.   .76
     × 8.4
     304
     608
     6.384

11.   .87
     × .69
     783
     522
     .6003

12.   9.5
     ×.07
     .665

13.   .407
     × .06
     .02442

14.   .92
     × .07
     .0644

15. a. .6 × 100 = .60. = 60
     b. 9.2 × 10 = 9.2. = 92
     c. 3.75 × 1,000 = 3.750.
       = 3,750

16.   44.8
     × 1.2
     896
     448
     53.76

17.   84.7
     × .68
     6776
     5082
     57.596

18.   9.02
     × .38
     7216
     2706
     3.4276

19. $39.43
     × 87
     27601
     31544
     $3,430.41

20. $1.29
     × 2.7
     903
     258
     $3.483 = $3.48

21.   .5
     × 3
     1.5 cups

22.   .015
      × 150
      ‾‾‾‾
      750
     15
    ‾‾‾‾‾‾
    2.250̸ = 2.25 in.

23.   18
    × .6
    ‾‾‾‾
    10.8 mi.

In counting up your answers, remember that there were 25 separate answers on this test.

A Score of	Means That You
23–25	Did very well. You can move on to Chapter 13.
20–22	Know this material except for a few points. Read the sections about the ones you missed.
16–19	Need to check carefully the sections you missed.
0–15	Need to review the chapter again to refresh your memory and improve your skills.

Questions	Are Covered in Section
1–3, 5, 6, 8, 11, 19	12.1
4, 7, 9–14, 16–18	12.2
15	12.3
20–25	12.4

ANSWERS FOR CHAPTER 12

PRETEST 12

1.	.8	2.	6.9	3.	5.26	4.	.4
5.	8.06	6.	.394	7.	.092	8.	2.16
9.	8.28	10.	5.628	11.	.7488	12.	.837
13.	.04224	14.	.0203	15.	a. 40	16.	42.38
					b. 12		
					c. 3,840		
17.	33.116	18.	3.9298	19.	$4,612.14		
20.	$1.8375 = $1.84	21.	1.5 lb.	22.	2 in.	23.	19.2 mi.

PRACTICE EXERCISE 121

1.	10.00 = 10	2.	12.00 = 12	3.	2.0 = 2	4.	18.48

5.	3.150 = 3.15	6.	.4	7.	2.50 = 2.5	8.	9.5
9.	.7245	10.	.36	11.	62.5	12.	21
13.	174	14.	2	15.	289.92	16.	4.8
17.	.12	18.	$100.66	19.	$3,218.60	20.	$500.00

PRACTICE EXERCISE 122

1.	1.848	2.	3.150 = 3.15	3.	.07245
4.	.036	5.	2.8992	6.	.480 = .48
7.	.0012	8.	$1.0066 = $1.01	9.	$324.786 = $324.79
10.	$68.750 = $68.75	11.	1.384	12.	134.712
13.	5.676	14.	.5024	15.	.00037
16.	$2.9136 = $2.91	17.	$22.467 = $22.47	18.	$176.000 = $176.00
19.	$812.592 = $812.59	20.	$3.51750 = $3.52		

PRACTICE EXERCISE 123

1.	42. = 42	2.	1,362.5	3.	14. = 14	4.	2400. = 2400
5.	2.47	6.	314.16	7.	6,200. = 6,200	8.	4. = 4
9.	$12,500. = $12,500			10.	$34.7 = $34.70		

PRACTICE EXERCISE 124

1.
$$\begin{array}{r} \$2.39 \\ \times\ \ 3.5 \\ \hline 1195 \\ 717 \\ \hline \$8.365 = \$8.37 \end{array}$$

2.
$$\begin{array}{r} \$2.97 \\ \times\ \ \ \ 52 \\ \hline 594 \\ 1485 \\ \hline \$154.44 \end{array}$$

3.
$$\begin{array}{r} .641 \\ \times\ \ \ 12 \\ \hline 1282 \\ 641 \\ \hline 7.692\ \text{in.} \end{array}$$

4.
$$\begin{array}{r} 22.6 \\ \times\ \ 11.8 \\ \hline 1808 \\ 226 \\ 226 \\ \hline 266.68\ \text{mi.} \end{array}$$

5. $1.28 × 1,000 = $1,280

6.
$$\begin{array}{r} \$48.45 \\ \times\ \ \ \ .07 \\ \hline \$3.3915 = \$3.39 \end{array}$$

7.
$$\begin{array}{r} \$.75 \\ \times\ \ \ \ 6 \\ \hline \$4.50 \end{array}$$

8.
$$\begin{array}{r} 31.37 \\ \times\ \ 26.8 \\ \hline 25096 \\ 18822 \\ 6274 \\ \hline 840.716\ \text{sq. ft.} \end{array}$$

9.
$$\begin{array}{r} 3.14 \\ \times\ \ \ 14 \\ \hline 1256 \\ 314 \\ \hline 43.96\ \text{in.} \end{array}$$

10.
$$\begin{array}{r} \$1.49 \\ \times\ \ \ 3.2 \\ \hline 298 \\ 447 \\ \hline \$4.678 = \$4.77 \end{array}$$

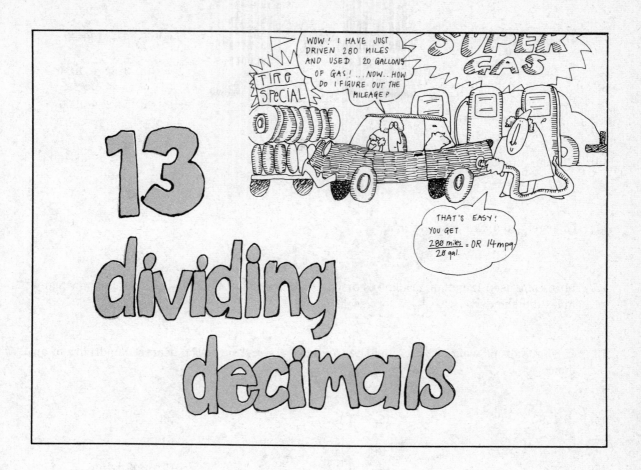

The topic covered in this chapter has various applications in our lives. You buy gasoline for your car, travel a given distance, and want to know how many miles you have traveled for each gallon of gasoline purchased. This is a problem in dividing decimals. You are offered a job with a yearly salary and want to know how many dollars and cents you will earn monthly or weekly. This is another example of dividing decimals.

Your knowledge of this topic can be checked with the pretest which follows.

See What You Know And Remember — Pretest 13

Do as many of these problems as you can. Try hard for a high score. Write each answer in the space provided.

1. $4 \overline{) .84}$

2. $3 \overline{) 9.3}$

3. $2 \overline{) 1.2}$

4. $8 \overline{) 40.8}$

5. $4 \overline{) .16}$

6. $3 \overline{) 3.009}$

7. $4\overline{)\,.084}$ 8. $6\overline{)\,.042}$ 9. a. $8 \div 100 =$

b. $1.32 \div 10 =$

c. $54.2 \div 1000 =$

10. Change $\dfrac{4}{5}$ to a decimal fraction.

11. Change $\dfrac{4}{9}$ to a two-place decimal.

12. Mrs. Ling used 12 gal. of gasoline to drive a distance of 146.4 mi. On the average, how many miles did she drive with 1 gal. of gasoline?

13. If 36 gumdrops weigh 11 oz., find the weight of a gumdrop to the nearest hundredth of an ounce.

14. $.3\overline{)\,.9}$ 15. $.5\overline{)\,.25}$ 16. $.7\overline{)\,4.9}$

17. $.8\overline{)\,64}$ 18. $.6\overline{)\,.72}$ 19. $.03\overline{)\,15.6}$

20. $.4\overline{)\,.002}$ 21. $.73\overline{)\,87.6}$ 22. $7.2\overline{)\,17.28}$

23. Divide 53,238 by .76.

24. Tomatoes sell for $.59 per lb. How many pounds did Mrs. Romero buy if she paid $3.54 for her tomatoes?

25. Mrs. Woo purchased $2\dfrac{2}{3}$ yd. of material for $7.92. What was the cost of 1 yd. of material?

26. If 2 lb. of potatoes cost $.23, how many pounds can be bought for $1.38?

27. Jane worked a year earning $12,194. How much did she earn weekly? (52 weeks = 1 year)

28. Find the average of 56, 63, 72, 45, 79, and 62 to the nearest tenth.

29. Change $.06\frac{1}{4}$ to a decimal.

30. Change $.87\frac{1}{2}$ to a fraction in its simplest form.

Now turn to the end of the chapter to check your answers. Add up all that you had correct. Count by the number of separate answers, not by the number of questions. In this pretest there were 30 questions, but 32 separate answers.

A Score of	Means That You
29–32	Did very well. You can proceed to Chapter 14 in Volume Three.
26–28	Know the material except for a few points. Read the sections about the ones you missed.
21–25	Need to check carefully on the sections you missed.
0–20	Need to work with this chapter to refresh your memory and improve your skills.

Questions	Are Covered in Section
1–4	13.1
5–8	13.2
10, 11	13.3
14–23	13.4
12, 13, 24, 26–28	13.5
9	13.6
30	13.7
29	13.8
25	13.9

13.1 Dividing A Decimal By A Whole Number

Example

If .8 lb. of cashew nuts are distributed among 4 people, how many pounds does each one get?

Solution

You must divide .8 by 4 or

$$4\overline{)\,.8\,}$$

to find the solution to this problem. Let's use our knowledge of fractions to discover the method of dividing a decimal by a whole number.

$$.8 \div 4 =$$

$$\frac{8}{10} \div \frac{4}{1} =$$

$$\frac{\overset{2}{\cancel{8}}}{10} \times \frac{1}{\underset{1}{\cancel{4}}} = \frac{2 \times 1}{10 \times 1} = \frac{2}{10} = .2$$

Thus: $4\overset{.2}{\overline{)\,.8\,}}$.

This example illustrates that you divide as in whole numbers and then place the decimal point in the quotient directly above the decimal point in the dividend. See the following example:

Example

Divide: $2.8 \div 4$.

Solution

Problem	Step 1	Step 2
$4\overline{)\,2.8\,}$	$4\overset{7}{\overline{)\,2.8\,}}$	$4\overset{.7}{\overline{)\,2.8\,}}$
	Divide: $28 \div 4 = 7$	Place the decimal point in the quotient (7) above the decimal point in the dividend.

Thus: $4\overset{.7}{\overline{)\,2.8\,}}$.

Practice Exercise 125

Divide.

1. $3 \overline{)\ .9}$

2. $6 \overline{)\ 26.4}$

3. $7 \overline{)\ 4.97}$

4. $5 \overline{)\ 4.15}$

5. $2 \overline{)\ .394}$

6. $4 \overline{)\ .92}$

7. $8 \overline{)\ 2.16}$

8. $9 \overline{)\ 67.5}$

9. $5 \overline{)\ .915}$

13.2   Zero As A Place Holder In The Quotient

Zero is used to hold a place between the decimal point and the first digit in the quotient. Sometimes there may be more than one zero. See the two examples which follow.

Example

Find the quotient of .16 divided by 16.

Solution

Problem	Step 1	Step 2
$16 \overline{)\ .16}$	$16 \overline{)\ .16}^{\;1}$	$16 \overline{)\ .16}^{\;.01}$
	Divide: $16 \div 16 = 1$ Place the 1 above the 6 in the dividend.	Place the decimal point in the quotient above the decimal point in the dividend. *Remember* to include the zero between the decimal point and the 1 as a place holder.

Thus: $16 \overline{)\ .16}^{\;.01}$.

And another —

Example

$$5 \overline{)\ .025}$$

Solution

Problem	Step 1	Step 2
$5 \overline{)\ .025}$	$\begin{array}{r} 5 \\ 5 \overline{)\ .025} \end{array}$	$\begin{array}{r} .005 \\ 5 \overline{)\ .025} \end{array}$
	Divide: $25 \div 5 = 5$ Place the 5 in the quotient above the dividend.	Place the decimal point in the quotient above the decimal point in the dividend. Include two zeros between the decimal point and the 5 as place holders.

Thus: $\begin{array}{r} .005 \\ 5 \overline{)\ .025} \end{array}$.

Practice Exercise 126

Divide. Remember to add zeros as place holders.

1. $3 \overline{)\ .09}$

2. $5 \overline{)\ .45}$

3. $9 \overline{)\ .081}$

4. $7 \overline{)\ .49}$

5. $6 \overline{)\ .0042}$

6. $4 \overline{)\ .052}$

7. $2 \overline{)\ .0034}$

8. $7 \overline{)\ .518}$

9. $6 \overline{)\ .036}$

10. $8 \overline{)\ .6464}$

11. $5 \overline{)\ .2715}$

12. $8 \overline{)\ .0736}$

13.3   Changing A Fraction Into A Decimal

A.   With No Remainder

Example

Juan decides to divide a gift of $3 among his 4 children. How much money does each one receive?

Solution

Juan knows that he must divide 3 into 4 equal parts. This means that each child will get $\frac{3}{4}$ of a dollar. The decimal fraction $.75 represents $\frac{3}{4}$ of a dollar.

One meaning of a fraction is that the line between the terms of the fraction means that the numerator is being divided by the denominator. In this example, $3 \div 4$ or $4\,)\,\overline{3}$. Look at the solution:

Problem	Step 1	Step 2	Step 3
$4\,)\,\overline{3}$	$4\,)\,\overline{3.}$	$4\,)\,\overline{3.00}$	$\begin{array}{r} .75 \\ 4\,)\,\overline{3.00} \end{array}$
	The decimal point is placed to the right of the whole number 3.	Since 4 cannot divide evenly into 3 add as many zeros to the right of the decimal point as you wish. Recall: 3 = 3. = 3.0 = 3.00, etc.	Place the decimal point in the quotient. Divide: $300 \div 4 = 75$

Thus: $\frac{3}{4}$ = .75 .

Practice Exercise 127

Find the decimal equivalent for each of these fractions by dividing the numerator by the denominator. Add zeros to the dividend until the problem works out evenly.

1. $\frac{2}{5}$ 2. $\frac{1}{4}$ 3. $\frac{1}{2}$

4. $\frac{1}{8}$ 5. $\frac{8}{10}$ 6. $\frac{5}{8}$

B.   With The Remainder Rounded Off

Suppose you divide and divide and divide and the problem never works out evenly. What do you do? Instructions to these examples indicate how far you should divide. Look at this example:

Example

Express the fraction $\frac{2}{3}$ as a decimal correct to the nearest hundredth.

Solution

The instructions in the example tell you that you must round off the answer to the nearest hundredth. This means that you will have a two-place decimal as your answer. Since you must round off your decimal to the hundredth, you must know the value of the digit in the thousandths place so you can round off properly. Look how the following example is worked out:

Problem	Step 1	Step 2	Step 3
$3\,\overline{)\,2}$	$3\,\overline{)\,2.000}$	$\overset{.666}{3\,\overline{)\,2.000}}$	$.666 = .67$
	Place the decimal point to the right of 2 and add 3 zeros (thousandths place): $2 = 2.000$	Divide. Find the quotient to one more place than the question calls for.	Round off to the nearest hundredth. Since the digit in the thousandths place (6) is more than 5, increase the digit in the hundredths place by 1: $.666 = .67$

Thus: $\frac{2}{3} = .67$ to the nearest hundredth.

And another —

Example

Find the quotient of $1.10 divided by 3 correct to the nearest cent.

Solution

Problem	Step 1	Step 2	Step 3
3) 1.10	3) 1.100	366 3) 1.100 9 —— 20 18 —— 20 18 ——	$.366 = $.37
	Add 1 zero to the dividend: 1.10 = 1.100	Divide. Find the quotient to the thousandths place: .366	Round off to the nearest cent (hundredths). $.366 = $.37

Thus: $1.10 ÷ 3 = $.37 to the nearest cent.

Practice Exercise 128

Find the decimal equivalent to the nearest tenth of:

1. $\frac{1}{3}$　　　　　2. $\frac{5}{8}$　　　　　3. $\frac{2}{7}$　　　　　4. $\frac{1}{9}$

Find the decimal equivalent to the nearest hundredth of:

5. $\frac{4}{9}$　　　　　6. $\frac{7}{8}$　　　　　7. $\frac{2}{7}$　　　　　8. $\frac{5}{6}$

9. Find correct to the nearest tenth:　4) 3.5

10. If a 6-ft. board is cut into 9 equal parts, find the length of each part correct to the nearest hundredth of a foot.

13.4   Dividing A Decimal By A Decimal

Now that you have seen that dividing a decimal by a whole number was hardly more difficult than ordinary division, you are ready to tackle division of a decimal by a decimal. Look at this example.

Example

A bag of candy weighs .6 lb. Into how many bags weighing .3 lb. each can it be divided?

Solution

The problem becomes .6 ÷ .3 = ? Let's use fractions to find the solution before tackling the problem of dividing a decimal by a decimal.

$$.6 \div .3 = \frac{.6}{.3}$$

$$= \frac{.6 \times 10}{.3 \times 10}$$

$$= \frac{6}{3} = 2$$

As you see, the divisor (.3) was made into a whole number by multiplying by 10. You must multiply both terms of a fraction by the same number to maintain an equivalent fraction, so multiply the numerator (.6) by 10 also.

Since the divisor (.3) was multiplied by 10, the result is the whole number 3. You must multiply the dividend (.6) by 10 also to maintain an equivalent fraction. Since the problem is written in fractional form, it is easier to use the terms numerator and denominator.

Recall: $\dfrac{\text{numerator}}{\text{denominator}} = \text{quotient}$

Thus you see that there is no special method for dividing a decimal by a decimal. All you have to do is change the divisor into a whole number and proceed as you did earlier in the chapter. Look at the following solution to this same example:

Problem	Step 1	Step 2
.3) .6	.3.) .6.	2. = 2 3) 6.
	Multiply the divisor by 10, 100, 1,000, etc., to make it a whole number. In this case, multiply by 10: .3 × 10 = 3. Multiply the dividend also by 10: .6 × 10 = 6 Place the decimal point in quotient directly above its new position in the dividend.	Divide the whole number (3) into the dividend (6): 6 ÷ 3 = 2

Thus: .6 ÷ .3 = 2.

And another —

Example

.04) .008

Solution

Be sure to change the divisor .04 into a whole number before dividing. Here is how:

Problem	Step 1	Step 2
.04) .008	.04.) .00.8	.2 4) .8
	Multiply the divisor (.04) by 100 to make it a whole number: .04 × 100 = .04. = 4 Multiply the dividend (.008) also by 100: .008 × 100 = .00.8 = .8 Place a decimal point in the quotient.	Divide: .8 ÷ 4 = .2

Thus: .008 ÷ .04 = .2 .

Practice Exercise 129

Divide. The decimal point has been left out of the quotient. Place the decimal point where it belongs. Put in zeros when necessary.

1. $\dfrac{3}{.3\,)\,.9}$ 2. $\dfrac{3}{.3\,)\,.09}$ 3. $\dfrac{3}{.03\,)\,.009}$ 4. $\dfrac{3}{.3\,)\,.009}$

5. $\dfrac{5}{.5\,)\,.025}$ 6. $\dfrac{5}{.05\,)\,.25}$ 7. $\dfrac{5}{.5\,)\,2.5}$ 8. $\dfrac{5}{.005\,)\,.0025}$

9. Divide: $.7342 \div .002$.

10. Find the quotient of $.034 \div .008$.

13.5   Zero As A Place Holder In The Dividend

Before dividing, it might be necessary to use zero as a place holder in the dividend.

Example

If a bus ride costs $.25, how many rides can you take for $6?

Solution

Problem	Step 1	Step 2
$.25\,)\,6$	$.25.\,)\,6.00.$	$\dfrac{24}{25\,)\,600}$
	Multiply the divisor (.25) by 100 to make it a whole number: $.25 \times 100 = 25$ Place the decimal point to the right of 6 in the dividend and multiply by 100: $6 \times 100 = 600$	Divide: $600 \div 25 = 24$

Thus: $\$6 \div \$.25 = 24$ rides.

210

And another —

Example

2.5 ÷ .005 =

Solution

Problem	Step 1	Step 2
.005 $\overline{)\,2.5}$	.005. $\overline{)\,2.500.}$	$\overset{500}{5\,\overline{)\,2,500}}$
	Multiply the divisor (.005) by 1,000: .005 × 1,000 = 5 Multiply the dividend 2.5 also by 1,000: 2.5 × 1,000 = 2.500. = 2,500	Divide: 2,500 ÷ 5 = 500

Thus: 2.5 ÷ .005 = 500.

Practice Exercise 130

The answers to the following problems have been completed except for the placement of the decimal point in the quotient. Place the decimal point where it belongs. Annex zeros when necessary.

1. $\overset{2\ 5}{25\,\overline{)\,62.5}}$
2. $\overset{25}{.25\,\overline{)\,6.25}}$
3. $\overset{25}{2.5\,\overline{)\,625}}$
4. $\overset{25}{.025\,\overline{)\,6.25}}$

5. $\overset{2\ 5}{.25\,\overline{)\,62.5}}$
6. $\overset{25}{2.5\,\overline{)\,.0625}}$
7. $\overset{2\ 5}{.025\,\overline{)\,62.5}}$
8. $\overset{25}{.25\,\overline{)\,625}}$

9. $\overset{25}{.0025\,\overline{)\,.625}}$
10. $\overset{25}{.25\,\overline{)\,.00625}}$

In this next group of problems, divide and place the decimal point in the quotient.

11. $5.6\,\overline{)\,4.592}$
12. $.81\,\overline{)\,429.3}$
13. $.98\,\overline{)\,.8526}$
14. $4.6\,\overline{)\,128.8}$

15. Find to the nearest tenth the quotient of 6.5 ÷ .261.

13.6 Word Problems Requiring The Division Of Decimals

Example

Sonia, who works for the A.B.C. Dress Company as a bookkeeper, received the following bill in the mail.

M.J.R. FABRIC CORPORATION
Houston, Texas

To: A.B.C. Dress Company
      Los Angeles, California

Quantity	Description	Amount
42 yd.	No. 123	$ 83.58
26 yd.	No. 710	93.34
104 yd.	No. 62C	186.16
	Total	$363.08

She has to check the price per yard of each item purchased and the total amount before she can pay the bill.

Solution

You divide each amount by the quantity received to find the price per yard or

$$\frac{\text{amount}}{\text{quantity}} = \text{price per yard.}$$

This means that each of these separate division examples must be completed as shown below.

```
      $ 1.99              $ 3.59               $   1.79
 42 ) $83.58         26 ) $93.34         104 ) $186.16
      42                  78                   104
     ────                ────                 ────
      41 5                15 3                 82 1
      37 8                13 0                 72 8
     ────                ────                 ────
       3 78               2 34                  9 36
       3 78               2 34                  9 36
     ────                ────                 ────
```

212

Thus:

$$42 \text{ yd. at } \$1.99 \text{ per yd.} = \$\ \ 83.58$$
$$26 \text{ yd. at } \$3.59 \text{ per yd.} = \$\ \ 93.34$$
$$104 \text{ yd. at } \$1.79 \text{ per yd.} = \$186.16$$
$$\$83.58 + \$93.34 + \$186.16 = \$363.08 \text{ Total}$$

Practice Exercise 131

1. Tomatoes sell for $.69 a pound. How many pounds can be purchased for $3.45?

2. If $60 is to be divided equally among 80 people, what decimal fraction of a dollar does each person receive?

3. A board 12.8 ft. long is to be cut into 4 equal lengths. Discounting the loss of material for each cut, what will be the length of each piece?

4. A parents' association purchased a quantity of candy to sell at the annual school bazaar. If the total cost of the 450 boxes of candy was $373.50, what was the price for each box?

5. Find the average price for the following 5 items if their individual prices are $.65, $1.28, $.25, $.98, and $1.19.

6. If the area of a rectangular plot of ground is 263.22 sq. in. and the length measures 21.4 in., find the measure of its width.

7. Find the number of miles a car travels on 1 gal. of gasoline if it requires 11.7 gal. to travel 289.2 mi. Round off the answer to the nearest tenth.

8. Mr. Gonzalez drove 180.6 mi. in 3.8 hr. Find his average rate of speed to the nearest tenth of a mile (speed = distance ÷ time).

9. A strip of metal 16.7 in. long is to be cut into 5 equal parts. What is the length of each part? (Allow nothing for the cut of the shears.)

10. A machine performs an operation every .8 sec. How many operations does it perform after working 28.8 sec.?

11. Mr. Kobetts purchased gasoline for his automobile for $8.28. If gasoline costs $.587 per gallon, how many gallons did he purchase? The answer should be rounded off to the nearest tenth of a gallon.

12. The St. Louis baseball team won 63 games and lost 72 games.

   a. Express the ratio of the number of games won divided by the total number of games played.

   b. Find the quotient to the nearest thousandth.

13. Apples sell for $.29 per pound. How many pounds did Mrs. Cappelletto buy if she paid $2.61 for her apples?

13.7 Special Divisors Of 10, 100, 1,000, Etc.

Example

A subscription to a sports magazine read as follows:

40 issues for $3.40
100 issues for $8.00

Which of the two choices would you make if you wanted to spend the least amount of money for each copy?

Solution

Divide the cost by the number of issues to find the price of one copy. Therefore:

```
              $ .085                                        $ .08
        40 ) $3.400            and                100 ) $8.00
              3 20                                          8 00
            ─────                                        ─────
               200
               200
            ─────
```

You were probably correct; it is cheaper to purchase 100 issues rather than 40 issues. For 40 issues you pay 8½¢ per copy while for 100 issues you pay only 8¢ a copy, representing a savings of ½¢ per copy.

In Chapter 12 you learned to multiply by the special numbers of 10, 100, 1,000, etc. You see you must divide by these numbers too. Let's look for and learn a short method for dividing by them as you did in multiplication. To discover the short cut, look at the following examples:

1. $8 \div 10 =$ 2. $8 \div 100 =$ 3. $8 \div 1,000 =$
```
         .8                      .08                        .008
   10 ) 8.0                100 ) 8.00            1,000 ) 8.000
```

When we divide by 10, the decimal point in the dividend (8.∅) moves *one place to the left* so that its position in the quotient is now .8 .

When we divide by 100, the decimal point in the dividend (8.∅∅) moves *two places to the left* so that its position in the quotient is now .08 .

When we divide by 1,000, the decimal point in the dividend (8.∅∅∅) moves *three places to the left* so that its position in the quotient is now .008 .

Practice Exercise 132

The quotients in problems 1–12 have been completed but the decimal point has been left out of the answer. Place the decimal point where it belongs. Put in zeros when necessary.

1. $42 \div 10 = 42$

2. $1,362.5 \div 100 = 13625$

3. $14 \div 10 = 14$

4. $2,400 = 1,000 = 24$

5. $2.47 \div 10 = 247$

6. $.456 \div 10 = 456$

7. $314.16 \div 100 = 31416$

8. $6,200 \div 10,000 = 62$

9. $4 \div 10 = 4$

10. $\$125.00 \div 100 = \$125$

11. $\$34.70 \div 10 = \$347$

12. $.456 \div 100 = 456$

215

13.8 Changing A Decimal Into A Fraction

A few pages back you learned to change a fraction into a decimal. Now look at the opposite operation, changing a decimal into a fraction.

Example

If two cans of corn cost $.25, how many cans can you purchase for $2?

Solution

If two cans of corn cost $.25, then $$.12\frac{1}{2}$$ is the cost of one can. To find the number of cans you can purchase, perform the following division example:

$$\$2 \div \$.12\frac{1}{2} = \qquad \text{or} \qquad .12.\frac{1}{2}\,\big)\,\overline{2.00.}$$

This problem is both complicated and difficult since you haven't attempted to solve any problem which contained $.12\frac{1}{2}$ as a divisor. What do you do?

The decimal $.12\frac{1}{2}$ is read as twelve and one-half hundredths and is written

$$\frac{12\frac{1}{2}}{100}$$

as a fraction. Look at the following solution:

Problem	Step 1	Step 2	Step 3	Step 4
$\dfrac{12\frac{1}{2}}{100}$	$12\frac{1}{2} \div 100$	$\dfrac{25}{2} \div \dfrac{100}{1}$	$\dfrac{\overset{1}{\cancel{25}}}{2} \times \dfrac{1}{\underset{4}{\cancel{100}}} =$	$\dfrac{1 \times 1}{2 \times 4} = \dfrac{1}{8}$
	Rewrite in this form.	Change the mixed number $12\frac{1}{2}$ to $\frac{25}{2}$.	Invert the divisor $\dfrac{100}{1}$ and cancel.	Multiply.

Thus: $.12\frac{1}{2} = \frac{1}{8}$.

Therefore:
$$\$2 \div \$.12\frac{1}{2} =$$
$$2 \div \frac{1}{8} =$$
$$\frac{2}{1} \times \frac{8}{1} = 16 \text{ cans}$$

The equality of $.12\frac{1}{2} = \frac{1}{8}$ resulted in a simple division example. Sometimes it is more convenient to use the fractional equivalent of a decimal when doing computations. One problem was just demonstrated. Look at the following example.

Example

Change .025 to a fraction in its simplest form.

Solution

Problem	Step 1	Step 2
$.025 =$	$\dfrac{25}{1,000}$	$\dfrac{25 \div 25}{1,000 \div 25} = \dfrac{1}{40}$
	Write the decimal as a fraction: $.025 = 25$ thousandths	Simplify the fraction: $\dfrac{25}{1,000} = \dfrac{1}{40}$

Thus: $.025 = \dfrac{1}{40}$.

And another —

Example

Change $.33\frac{1}{3}$ to a fraction in its simplest form.

Solution

Problem	Step 1	Step 2	Step 3	Step 4
$.33\frac{1}{3} =$	$\dfrac{33\frac{1}{3}}{100}$	$33\frac{1}{3} \div 100 =$	$\dfrac{100}{3} \div \dfrac{100}{1} =$	$\dfrac{\overset{1}{\cancel{100}}}{3} \times \dfrac{1}{\underset{1}{\cancel{100}}} = \dfrac{1}{3}$
	Write as a fraction.	Rewrite in this form.	Change $33\frac{1}{3}$ to an improper fraction.	Multiply.

Thus: $.33\frac{1}{3} = \dfrac{1}{3}$.

You are now able to convert any decimal to a fraction or any fraction to a decimal. Choosing one rather than another usually depends upon the nature of the problem. There are some fractions and decimals which are used more frequently than others. These should be familiar to you and you should commit them to memory. Many of them are included in the next practice exercise.

Practice Exercise 133

In the following table certain values are missing. Find the fraction or decimal equivalent and place it in the table.

	Fraction	Decimal Equivalent			Fraction	Decimal Equivalent
1.	$\frac{1}{2}$	.50	11.	$\frac{4}{5}$	
2.	$\frac{1}{3}$	$.33\frac{1}{3}$	12.		$.83\frac{1}{3}$
3.	$\frac{1}{4}$	.25	13.		$.62\frac{1}{2}$
4.		.20	14.		$.87\frac{1}{2}$
5.	$\frac{1}{6}$		15.	$\frac{7}{10}$	
6.		$.12\frac{1}{2}$	16.		.9
7.	$\frac{1}{10}$		17.		$.66\frac{2}{3}$
8.		.4	18.	$\frac{3}{4}$	
9.	$\frac{3}{5}$		19.	$\frac{3}{8}$	
10.		.3	20.		$.42\frac{6}{7}$

13.9   Rewriting Decimals With Fractions

An area of difficulty is the decimal and fraction combination like $.12\frac{1}{2}$. You saw that it can be changed into the fraction $\frac{1}{8}$. Can it be written as a decimal number without a fraction?

Look at the decimal number .125. What is its fractional value? It is $\dfrac{125}{1,000} = \dfrac{1}{8}$. Thus: $.12\dfrac{1}{2} = .125$.

Example

Is it true that $.06\dfrac{1}{4} = .0625$?

Solution

$$.06\dfrac{1}{4} = \dfrac{6\dfrac{1}{4}}{100} = 6\dfrac{1}{4} \div 100 \qquad\qquad .0625 = \dfrac{625}{10,000}$$

$$= \dfrac{25}{4} \div \dfrac{100}{1} \qquad\qquad\qquad = \dfrac{625 \div 625}{10,000 \div 625}$$

$$= \dfrac{\overset{1}{\cancel{25}}}{4} \times \dfrac{1}{\underset{4}{\cancel{100}}} \qquad\qquad\qquad = \dfrac{1}{16}$$

$$= \dfrac{1}{16}$$

Thus: $.06\dfrac{1}{4} = .0625$.

To rewrite a decimal containing a fraction like $\dfrac{1}{2}$, $\dfrac{1}{4}$, etc., replace the fraction with its decimal equivalent, $\dfrac{1}{2} = .5$, $\dfrac{1}{4} = .25$, etc. Write the decimal equivalent to the right of the original decimal but do not include another decimal point. Look at the illustrations below.

1. $.07\dfrac{1}{4} = .0725$ 2. $.02\dfrac{1}{2} = .025$ 3. $3.42\dfrac{1}{8} = 3.42125$

     since $\dfrac{1}{4} = .25$ since $\dfrac{1}{2} = .5$ since $\dfrac{1}{8} = .125$

Practice Exercise 134

Change each decimal to a common fraction or an improper fraction in its simplest form.

1. $.6 =$ 2. $.04 =$ 3. $.375 =$

4. $.07\frac{3}{4} =$ 5. $.222 =$ 6. $.15 =$

7. $1.2 =$ 8. $.80 =$ 9. $.62\frac{1}{2} =$

10. $.83\frac{1}{3} =$

11. If 3 cans of peas cost $1, then how many cans of peas can be purchased for $4?

12. If a towel requires $.66\frac{2}{3}$ yd. of material, then how much material would be needed to make 8 towels for the family?

13.10   Word Problems Combining Fractions And Decimals

Example

The cost of a dozen cans of soda is $1.20. What is the cost of $\frac{3}{4}$ of a dozen?

Solution

To determine the cost of the purchase find $\frac{3}{4} \times \$1.20$ using:

Method 1
(all fractions)

$\$1.20 = 1\frac{20}{100} = 1\frac{1}{5}$

$\frac{3}{4} \times 1\frac{1}{5} =$

$\frac{3}{\underset{2}{4}} \times \overset{3}{\cancel{6}}{5} = \frac{9}{10} = \$.90$

Method 2
(all decimals)

$\frac{3}{4} = .75$

$\begin{array}{r} \$1.20 \\ \times \quad .75 \\ \hline 600 \\ 840 \\ \hline \$.9000 = \$.90 \end{array}$

Method 3
(fractions and decimals)

$\frac{3}{4}$ of $\$1.20 =$

$\frac{3}{\underset{1}{4}} \times \frac{\overset{.30}{\cancel{\$1.20}}}{1} = \$.90$

220

And another —

Example

Yolanda bought $3\frac{1}{4}$ yd. of dress material at $3.96 per yd. How much did the material cost her?

Solution

One yard of material costs $3.96. To find the cost of $3\frac{1}{4}$ yards, multiply $3.96 by $3\frac{1}{4}$ using:

Method 1	Method 2	Method 3
(all fractions)	(all decimals)	(fractions and decimals)

Method 1 (all fractions)

$3\frac{1}{4} \times \$3.96 =$

$3\frac{1}{4} \times 3\frac{96}{100} =$

$\frac{13}{4} \times \frac{396}{100} = \frac{1287}{100}$

$= 12\frac{87}{100}$

$= \$12.87$

Method 2 (all decimals)

$3\frac{1}{4} = 3.25$ yd.

$$\begin{array}{r} \$3.96 \\ \times \quad 3.25 \\ \hline 1980 \\ 792 \\ 1188 \\ \hline \$12.8700 = \$12.87 \end{array}$$

Method 3 (fractions and decimals)

$3\frac{1}{4} \times \$3.96 =$

$$\frac{13}{\cancel{4}} \times \frac{\overset{.99}{\cancel{\$3.96}}}{1} = \$12.87$$
$$1$$

The methods just illustrated can be used as you desire. It is up to you to choose that method which best fits the problem and best suits you. You should be aware of each method and be able to do any one of them at any time.

Practice Exercise 135

Find the value of each of the following problems.

1. Find $\frac{1}{2}$ of $4.02.

2. $.04 \div \frac{4}{5} =$

3. $\$6.03 \times \frac{2}{3} =$

4. Find $\frac{3}{4}$ of .0056 in.

5. Find the quotient of $.62 \div \frac{5}{8}$ to the nearest hundredth.

6. If two pieces of chocolate cost $.13 then what is the cost of 17 pieces?

7. If a foot of wire costs $.04 $\frac{1}{4}$, find the number of feet of wire purchased for $1.02.

8. If a baseball player gets a hit .275 times for each of the 80 times he was at bat, then how many hits did he get?

9. Baseball standings are represented by a three-place decimal. This decimal is found by forming a fraction of the number of games won divided by the total number of games played. In the Cape League standings which follow, find the 3-place decimal for each team. Orleans has been completed for you.

	CAPE LEAGUE STANDINGS	
	W	L
Orleans	19	11
Wareham	20	12
Cotuit	17	16
Harwich	18	17
Chatham	17	18
Falmouth	13	19
Yarmouth	11	22

League	W	L	Fraction	Decimal
Orleans	19	11	$\frac{19}{30}$	$\frac{19}{30} = 30\overline{)19.0000}$ .6333 $= .6333 = .633$
Wareham				
Cotuit				
Harwich				

League	W	L	Fraction	Decimal
Chatham				
Falmouth				
Yarmouth				

10. If a $\frac{3}{8}$ in. rivet weighs .37 $\frac{1}{2}$ lb., how many rivets are there in a pile that weighs 67 $\frac{1}{2}$ lb.?

Review Of Important Ideas

Some of the most important ideas in Chapter 13 were:

 Dividing decimals is almost identical to dividing whole numbers.

 Division by a decimal is accomplished after the divisor is made into a whole number.

 To divide a decimal by a whole number —
1. Divide as you would with whole numbers.
2. Place the decimal point in the quotient directly above the decimal point in the dividend.

 To divide a decimal by a decimal —
1. Multiply the divisor by a multiple of 10 so that it becomes a whole number.
2. Multiply the dividend by the same number used to multiply the divisor.
3. Proceed as you did when dividing a decimal by a whole number.

To divide by the special numbers of 10, 100, 1,000, etc. —
1. Keep the same digits in the quotient as there are in the dividend.
2. Move the decimal point in the dividend as many places to the left as there are zeros in the divisor.

 To change a fraction into a decimal —
1. Place the decimal point in the dividend (fraction's numerator) at the extreme right of the number.
2. Annex as many zeros to the right of the decimal point in the dividend as desired.
3. Divide until there is no remainder, or to one decimal place more than the question calls for. Round off to the desired number of decimal places.

 To change a decimal to a fraction —
1. Write the numerator of the fraction having the same digits as the decimal without the decimal point.
2. Write the denominator as a multiple of 10, 100, 1,000, etc., according to the number of digits in the numerator.
3. Simplify the resulting fraction if possible.

 When doing computations with both decimals and fractions —
1. Change all the numbers to fractions and then proceed.
2. Change all the numbers to decimals and then proceed.
3. Do the computations without changing to the same type of terms.

Check What You Have Learned

The examples in this posttest will test your progress in dividing decimals. Try to get a satisfactory grade.

Posttest 13

1. $3\overline{)\,.63}$ 2. $4\overline{)\,8.4}$ 3. $3\overline{)\,1.2}$ 4. $7\overline{)\,35.7}$

5. $6\overline{)\,.18}$ 6. $7\overline{)\,7.007}$ 7. $2\overline{)\,.084}$ 8. $9\overline{)\,.081}$

9. a. $.6 \div 100 =$ b. $4.27 \div 10 =$ c. $63.8 \div 1,000 =$

10. Change $\frac{3}{5}$ to a decimal fraction.

11. Change $\frac{2}{7}$ to a two-place decimal.

12. Wai Hung used 13 gal. of gasoline to drive a distance of 158.6 mi. On the average, how many miles did she drive with 1 gal. of gasoline?

13. If 45 gumdrops weigh 14 oz., find the weight of a gumdrop to the nearest hundredth of an ounce.

14. $.4 \overline{)\ .8}$ 15. $.8 \overline{)\ .72}$ 16. $.6 \overline{)\ 4.2}$ 17. $.5 \overline{)\ 25}$

18. $.7 \overline{)\ .84}$ 19. $.04 \overline{)\ 16.8}$ 20. $.6 \overline{)\ .003}$ 21. $.87 \overline{)\ 104.4}$

22. $6.3 \overline{)\ 15.12}$

23. Divide 48,334.5 by .69 .

24. Potatoes sell for $.19 per lb. How many pounds did Mrs. Phillips buy if she paid $1.14 for her potatoes?

25. Diane paid $110.80 for $34\frac{5}{8}$ yd. of drapery material. What was the cost of 1 yd. of the material?

26. If 3 lb. of onions cost $.29, how many pounds can be bought for $.87?

27. Mary worked a year, earning $6,994. How much did she earn weekly? (52 weeks = 1 year)

28. Find the average of 38, 41, 35, 45, 49, and 37 to the nearest tenth.

29. Change $.08\frac{1}{5}$ to a decimal.

30. Change $.62\frac{1}{2}$ to a fraction in its simplest form.

1. $3\overline{).63}$ .21

2. $4\overline{)8.4}$ 2.1

3. $3\overline{)1.2}$ .4

4. $7\overline{)35.7}$ 5.1

5. $6\overline{).18}$ .03

6. $7\overline{)7.007}$ 1.001

7. $2\overline{).084}$ .042

8. $9\overline{).081}$ .009

9. a. .6 ÷ 100 = .00.6 = .006   b. 4.27 ÷ 10 = .4.27 = .427   c. 63.8 ÷ 1,000 = .063.8 = .0638

10. $\dfrac{3}{5} = 5\overline{)3.0}$ .6

11. $\dfrac{2}{7} = 7\overline{)2.00}$ .285 = .29

12.
```
        12.2 mi.
13 ) 158.6
     13
     ──
     28
     26
     ──
      2 6
      2 6
```

13.
```
              .3111 = .311 oz.
45 ) 14.0000
     13 5
     ────
       50
       45
       ──
       50
       45
       ──
       50
       45
       ──
        5
```

14. $.4\overline{).8}$ 2.

15. $.8\overline{).7.2}$ .9

16. $.6\overline{)4.2}$ 7.

17. $.5\overline{)25.0}$ 5 0.

18. $.7\overline{).8.4}$ 1.2

19. $.04\overline{)16.80.}$ 4 20.

20. $.6\overline{).0.030}$ .005

21.
```
        1 20.
.87 ) 104.40.
      87
      ──
      17 4
      17 4
```

22.
```
       2.4
6.3 ) 15.1 2
      12 6
      ────
       2 5 2
       2 5 2
```

23.
```
          70,050.
.69. ) 48,334.50.
       48 3
       ────
        34 5
        34 5
```

24. $.19\overline{)1.14.}$ 6. = 6 lb.

226

25. $110.80 \div 34\frac{5}{8} =$

$110.80 \div \frac{277}{8} =$

$\frac{\overset{.40}{\cancel{110.80}}}{\underset{1}{0}} \times \frac{8}{\cancel{277}} = \$3.20 \text{ per yd.}$

26.
$$\overset{3.}{\$.29.\,)\,\$.87.}$$
$3 \times 3 = 9 \text{ lb.}$

27.
$$\overset{\$\ 134.50}{52\,)\,\$6,994.0}$$
$5\ 2$
$\overline{}$
$1\ 79$
$1\ 56$
$\overline{}$
234
208
$\overline{}$
$26\ 0$
$26\ 0$
$\overline{}$

28. $\dfrac{38 + 41 + 35 + 45 + 49 + 37}{6}$

$= \dfrac{245}{6} = 40.8$

29. $.08\frac{1}{5} = .082$

30. $.62\frac{1}{2} = \dfrac{62\frac{1}{2}}{100} = 62\frac{1}{2} \div 100$

$= \dfrac{\overset{5}{\cancel{125}}}{2} \times \dfrac{1}{\underset{4}{\cancel{100}}} = \dfrac{5}{8}$

In counting up your answers, remember that there were 32 separate answers in this test.

A Score of	Means That You
29–32	Did very well. You can proceed to Chapter 14 in Volume Three.
26–28	Know this material except for a few points. Reread the sections about the ones you missed.
21–25	Need to check carefully the sections you missed.
0–20	Need to review the chapter again to refresh your memory and improve your skills.

Questions	Are Covered in Section
1–4	13.1
5–8	13.2
10, 11	13.3
14–23	13.4
12–13, 24, 26–28	13.5
9	13.6
30	13.7
29	13.8
25	13.9

ANSWERS FOR CHAPTER 13

PRETEST 13

1.	.21	2.	3.1	3.	.6
4.	5.1	5.	.04	6.	1.003
7.	.021	8.	.007		
9.	a. .08		b. .132		c. .0542
10.	.8	11.	.44	12.	12.2 mi.
13.	.31 oz.	14.	3	15.	.5
16.	7. = 7	17.	80. = 80	18.	1.2
19.	520	20.	.005	21.	120
22.	2.4	23.	70,050	24.	6 lb.
25.	$2.97	26.	12 lb.	27.	$234.50
28.	62.8	29.	.0625	30.	$\frac{7}{8}$

PRACTICE EXERCISE 125

1.	.3	2.	4.4	3.	.71
4.	.83	5.	.197	6.	.23
7.	.27	8.	7.5	9.	.183

PRACTICE EXERCISE 126

1.	.03	2.	.09	3.	.009
4.	.07	5.	.0007	6.	.013
7.	.0017	8.	.074	9.	.006
10.	.0808	11.	.0543	12.	.0092

PRACTICE EXERCISE 127

1.	.4	2.	.25	3.	.5
4.	.125	5.	.8	6.	.625

PRACTICE EXERCISE 128

1.	.3	2.	.6	3.	.3	4.	.1
5.	.44	6.	.88	7.	.29	8.	.83
9.	.9	10.	.67 ft.				

PRACTICE EXERCISE 129

1.	3	2.	.3	3.	.3	4.	.03
5.	.05	6.	5	7.	5	8.	.5
9.	367.1	10.	4.25				

PRACTICE EXERCISE 130

1. 2.5		**2.** 25		**3.** 250		**4.** 250
5. 250		**6.** .025		**7.** 2,500		**8.** 2,500
9. 250		**10.** .025		**11.** .82		**12.** 530
13. .87		**14.** 28		**15.** 24.9		

PRACTICE EXERCISE 131

1. $.69. \overline{)3.45.}$ → 5 lb.

2. $\frac{60}{80} = \frac{3}{4} = \$.75$

3. $4\overline{)12.8}$ → 3.2 ft.

4. $450\overline{)373.50}$ → $\$.83$

5. $\frac{.65 + 1.28 + .25 + .98 + 1.19}{5} = \frac{4.35}{5} = \$.87$

6. $\frac{263.22}{21.4} = 12.3$ in.

7. $11.7.\overline{)289.2.}$ → 24.7 mi.

8. 47.5 m.p.h.

9. $5\overline{)16.7}$ → 3.34 in.

10. $.8.\overline{)28.8.}$ → $36.$ operations

11. 14.1 gal.

12. a. $\frac{63}{135}$ b. .467

13. $.29.\overline{)2.61.}$ → 9 lb.

PRACTICE EXERCISE 132

1. 4.2		**2.** 13.625		**3.** 1.4		**4.** 2.4
5. .247		**6.** .0456		**7.** 3.1416		**8.** .62
9. .4		**10.** $1.25		**11.** $3.47		**12.** .00456

PRACTICE EXERCISE 133

4. $\frac{1}{5}$		**5.** $.16\frac{2}{3}$		**6.** $\frac{1}{8}$		**7.** .1	**8.** $\frac{2}{5}$
9. .6		**10.** $\frac{3}{10}$		**11.** .8		**12.** $\frac{5}{6}$	**13.** $\frac{5}{8}$
14. $\frac{7}{8}$		**15.** .7		**16.** $\frac{9}{10}$		**17.** $\frac{2}{3}$	**18.** .75
19. .375		**20.** $\frac{3}{7}$					

PRACTICE EXERCISE 134

1. $\frac{3}{5}$		**2.** $\frac{1}{25}$		**3.** $\frac{3}{8}$		**4.** $\frac{31}{400}$
5. $\frac{111}{500}$		**6.** $\frac{3}{20}$		**7.** $1\frac{1}{5}$		**8.** $\frac{4}{5}$
9. $\frac{5}{8}$		**10.** $\frac{5}{6}$		**11.** 12 cans		**12.** $5\frac{1}{3}$ yd.

1. $2.01
2. .032
3. $4.02
4. .0042 in.
5. .99
6. $1.11
7. 24 ft.
8. 22 hits

9.

	Fraction	Decimal
Wareham	$\frac{20}{32}$	.625
Cotuit	$\frac{17}{33}$	.515
Harwich	$\frac{18}{35}$	.514
Chatham	$\frac{17}{35}$	.486
Falmouth	$\frac{13}{32}$	.406
Yarmouth	$\frac{11}{33}$	.333

10. 180 rivets

You Are Almost There

This completes your study of decimals. You have seen how, where, and why decimals are used in everyday life. You have only scratched the surface in these chapters but we hope that this study will help you to use mathematics, to feel better about yourself as a person, or just to feel more secure when mathematics is used to explain a happening. Now you are ready for Volume Three.

Hold It!

Are your arithemetic skills improving? This test will allow you to find out.

1. $\frac{5}{8} \times 2\frac{3}{10} =$

2. $3.16 \times .42 =$

3. $.036 \div 12 =$

4. Change $\frac{2}{9}$ to a two-place decimal.

5. $1.6 + 12 + .483 =$

6. $14\frac{5}{8}$
   $+ 4\frac{1}{2}$
   _____

7. Change .025 to a common fraction in its simplest form.

8. The sales tax on an item is $4.292. What is this amount rounded off to the nearest cent?

9. If a runner sets a new world record of eight and eighty-seven hundredths seconds, how would this be written as a decimal numeral?

10. Find the quotient of 5.16 ÷ 1.2.

11. The cost of 18,800 sq. ft. of property is $5,500. What is the cost per sq. ft. correct to the nearest cent?

12. a. What is the cost of $2\frac{1}{2}$ lbs. of rump roast and a $2\frac{1}{4}$ lb. pork tenderloin at the following prices?

face rump roast	$1.49 lb.
pork tenderloin	1.59 lb.

   b. If you give the butcher a $10 bill, how much change do you get?

ANSWERS TO "HOLD IT!"

1. $1\frac{7}{16}$ 2. 1.3272   3. .003   4. .22   5. 14.083

6. $19\frac{1}{8}$ 7. $\frac{1}{40}$ 8. $4.29   9. 8.87 sec.   10. 4.3

11. $.29 per sq. ft.   12. a. $3.73   b. $10.00
                              + 3.58   − 7.31
                              ───────   ─────────
                               $7.31 $ 2.69